I ♥ My Cat!

The All-Around Guide to Choosing, Grooming, Raising and Caring for Your Feline Friend

WITH PHOTOGRAPHS BY LAURA MOSS

Woman'sDay

Copyright © 2010 Hachette Filipacchi Media U.S., Inc.

First published in 2010 in the United States of America by Filipacchi Publishing
1633 Broadway
New York, NY 10019

Woman's Day is a registered trademark of Hachette Filipacchi Media U.S., Inc.

Printed in China

Design: Patricia Fabricant
Editor: Lauren Kuczala
Production: Lynn Scaglione and Annie Andres

ISBN-13: 978-1-936297-00-9
Library of Congress Control Number: 2010921541

*If man could be crossed with
the cat it would improve man, but it would
deteriorate the cat.*

Mark Twain

The smallest feline is a masterpiece.

Leonardo da Vinci

CONTENTS

CATS 101

*A **kitten's cuteness** is the major reason people end up adopting one before they have researched how to choose, nurture, feed or care for their new pet. Just the sight of a kitten's adorable "pansy" face, tiny nose, bright eyes and itsy-bitsy tail, and often hearts make decisions that brains might not.*

Cats can live to be 20 years old or older, so adopting one is a serious commitment of time and money. It's hard to visualize the cons when you look into a gorgeous kitten face, but the reality is that kittens can be destructive and, if handled improperly, can scratch or bite. They need to be fed, their litter box needs to be cleaned frequently and they need routine health care. Kittens don't stay little fuzzballs, either. They grow up to be cats and lose their kittenish looks.

Still interested in adopting a kitten? The information in this book will help you to choose, raise, understand and care for your new furry pal.

Interesting Animal Facts: All About Cats

Cats are exceptionally warm animals. In fact, most people do not know that the average body temperature of a cat is 101.5°F, a few degrees higher than a human's own 98.6°F. Here are some other interesting facts that you likely never knew about cats.

- **Functions of whiskers** Have you ever wondered why cats have whiskers? The long hairs, which are flexible and about three times thicker than the rest of a cat's hair, have several purposes. Whiskers work as motion detectors and help cats find their way around. Whiskers are also indicative of a cat's mood, so be sure to pay attention to your kitty's whiskers. The less relaxed the whiskers are, the more likely your cat is to be frustrated or angry.

- **Largest and smallest cats** Many people may already know that tigers are the largest cat. However, here are some interesting animal facts that you may not know. Tigers can weigh up to 800 pounds, about eight times the size of a very large dog. The Singapura, which is a domestic breed, is the smallest type of cat. On average, these cats weigh about 5 pounds, which is no more than a teacup dog. The Ragdoll, which can weigh up to 20 pounds, is the largest domestic cat breed.

- **Their own "fingerprints"** Humans are not the only ones with personally unique features. One of the most interesting animal facts about both domestic and wild felines is that they each have something which is uniquely their own. It is their nose, which is made up of a pattern of different ridges. Believe it or not, no two cats have the same exact nose.

- **More bones and sleep than humans** Did you know that a cat has more bones than you do? Humans have 206 bones, whereas cats have 230. On average, cats sleep 16 hours a day, which is double the daily recommended amount of sleep for humans.

- **Cats and their senses** Kittens are born deaf. Their sense of hearing does not develop for a few weeks. Cats also have a very strong sense of smell, so keep in mind that the litter box smells stronger to them than it does to you.

- **Overpopulation of cats** A cat has the potential to give birth to 100 kittens in her lifetime. Litters usually consist of anywhere between one and eight kittens and may occur two to three times a year. Cats are twice as likely to end up in an animal shelter than dogs are. Be sure to have your cat spayed or neutered to prevent adding to overpopulation.

How to Pick the Purrfect Kitten for Your Family

You have your heart set on an adorable Persian. But the tabby next to her might be a better fit. One of the biggest mistakes when choosing a cat is picking purely on looks. Before the cuteness factor takes over completely, ask yourself:

Is My Family Ready?
Don't let your kids' pleas direct your decision. First, consider allergies. About 10 million Americans are allergic to cat dander—make sure your family isn't among them.

Second, think about other pets you have. Consult with your vet to see if a cat will be compatible with your dog or other pet. Also ask the vet about specific traits. Some breeds are more sociable (like Siamese) and some are more prone to health problems (such as kidney disease in Abyssinians).

Finally, be honest about upkeep. Like any pet, cats come with a price tag: food, litter, vet visits. Earmark a savings fund for your cat in case of an emergency ($200 or so), or consider pet insurance.

How Old Should It Be?
Most kittens are furballs of energy—they get into everything. But some of them are a bit less hyper (though not much!).

If you get a kitten that's at least 6 months old, you'll be able to tell if it's social and active or quiet and laid-back. If possible, bring the cat home for a day before you commit to see how it interacts with your family.

What Does a Healthy Cat or Kitten Look Like?

Just look at that gorgeous, healthy kitty curled up on your sofa. Everything from the swish of her tail to the shine in her eyes lets you know what a strong and healthy cat she is. You can judge a cat's health just by watching her behavior and examining her physical characteristics.

- **Behavior** No two cats are the same, but all healthy cats and kittens will be responsive to stimuli and curious about new things. Healthy cats will be playful and alert. They'll be able to move fluidly and show off their great balance. Healthy cats will also have good appetites and their stool should be soft-firm. A healthy cat's litter box should get daily deposits of solid and liquid wastes.

- **Eyes, ears and nose** Nothing speaks to cat health like clear, bright eyes. Their pupils should be reactive and the same size. The whites of their eyes should be truly white. Healthy cats' eyes will never be runny or weepy. Normally, seeing a cat's "third eyelid," the membrane in the inner corner of an eye, indicates poor health and is a warning sign that should not be ignored.

  A cat's ears should be clean and slightly pink. A small amount of ear wax is normal in a healthy cat, but there shouldn't be any excessive buildup or any sores, crusty bits or foul odor. Black, tarry splotches in her ear can indicate an infestation of ear mites.

  A healthy cat's nose can be dry or moist, but never wet. A runny nose is a sure sign of a sick kitty.

- **Mouth** A healthy cat's gums will be pink and clean. Very pale or red gums can be a sign something is wrong. Blue gums may mean there's an oxygen deficiency. A cat's teeth should be white and without visible tartar. Obviously, older cats have teeth that have grown darker through the years, but black-spotted teeth can indicate decay. Bad breath is often a sign of dental or gum disease.

- **Coat** A healthy cat's coat should be smooth and soft. There should not be bald patches, nor should the fur feel either oily or dry. A healthy, shiny coat is a sure sign your cat is eating a balanced diet and receiving proper nutrition.

  A cat's skin can range in color from pink to dark brown. It should be dandruff-free and supple. If you gently grasp some skin at the base of her neck and pull it up, the skin should quickly return to normal in a healthy cat. If it's slow to snap back, your cat may be dehydrated.

- **Body shape** Cats come in all shapes and sizes, but they should not be either heavy or thin. Two simple tests can help you determine if your cat is underweight or overweight. Look at your cat from above. You should be able to see an hourglass shape. A healthy cat's waist will always be visible. Next, feel her rib cage. You should be able to feel her ribs, but not see them.

Adopting the Perfect Pet

Getting a new pet is a big decision, and if you want to adopt one from a shelter, there are several factors to consider. Betsy Saul, cofounder of Petfinder.com, a website that lists pets available for adoption from shelters and rescue organizations, helped get 2.5 million pets adopted last year. Betsy, who owns 20 rescues (a calf, two goats, a sheep, two horses, two cats, two turtles, two guinea pigs, seven chickens and a 15-year-old dog named Jim, whom she found in Costa Rica), knows a thing or two about finding the perfect match.

1 **Get to know your shelter.** Some shelters and rescue groups have lengthy questionnaires, required home visits and expensive adoption fees. Others are animal control units—you pay $8 and you can walk away with a pet. Most shelters fall somewhere in between, charging an adoption fee of $100 to $200. "It can be anywhere from an intense relationship to no relationship with the shelter, so check out the policies, fees and attitude of the place you're going to before you make a commitment—because sometimes, it's almost like you're adopting the shelter, too," says Betsy. The choice is yours.

2 **Think about your lifestyle.** Do you come home from work and just want to veg? Or are you going to be ready for a run? "Most people have an image of the dog that they want, but sometimes our perceptions of our own life do not match reality," says Betsy. You might love the outdoors, but how often do you really go hiking? "That hiking dog who will only go hiking once a year will need more exercise than you can give him."

3 **Consider an older pet.** You may think you want a young dog or cat because the baby stage is so cute. But older animals can be much easier. "They've seen a lot, and when you spring them from a shelter, they love you because you're the person who got them out of there. And they're often already house-trained," says Betsy, who generally only adopts older animals. When it comes down to it, there is no perfect age. It's all about your own time commitment.

4 Consider the breed. Petfinder advocates mixed breeds—although 25 percent of the animals on the site are purebred.

5 Get away from the mayhem. "Remove the cat or dog from the chaos of the other animals, whether that's a private room for a cat or a long walk for a dog," says Betsy. That's when his or her real personality will begin to shine through.

6 Take the whole household into account. Have another cat at home? Go to Petfinder.com and watch their video on how to introduce two cats. For cats in general, the most successful adoptions happen in pairs. If you had a pet who recently passed away and have another one still at home, it's not a bad idea to get a new pet fairly quickly so the remaining pet has somewhere to direct its love. "It's not replacing your pet—it's honoring that amazing life by saving another one," says Betsy.

7 Foster to adopt. Many shelters and rescue groups have foster-to-adopt programs, where you take the animal into your home for a period of time. It's sort of like test-driving. It helps you assess whether this pet fits into your lifestyle and gives everyone time to adjust and get to know each other.

8 Give it time. A surprising number of animals are returned to the shelter, so know ahead of time that there will be an adjustment period. "You'll need a couple of weeks for integration, and at the 3- to 6-month mark, they'll realize that this is their home. I had a dog that didn't bark at all. I thought there might be something wrong with him, then one day, I was standing at the stairs and all of a sudden, he was talking about a squirrel. He'd finally gotten comfortable enough to be himself," says Betsy.

Essential Pet Supplies for New Cat Owners

You've decided to add a frisky feline to your family, but there's more to being a cat owner than simply bringing home the kitty. There are a few essential pet supplies new cat owners need to bring home, too.

- **Cat food and dishes** Cats need to eat, so make sure to provide the best cat food you can afford. If your new pet is a kitten, choose a quality kitten food for that tiny tummy. Don't forget a dish for the cat food and another for water. Avoid using plastic cat dishes, because plastics contain an odor that can be off-putting to your cat. Make sure clean, fresh water is available at all times.

- **Litter and litter box** It's a fact: Cats need their own "throne room," or at least their own "throne box." A litter box and cat litter are must-haves in a house with a cat. A quality scoopable litter (don't forget the scoop!) and a covered litter box are good choices for your new cat.

- **Toys** Whether you're bringing home a young kitten or an adult cat, most cats like to play. While cats generally spend the majority of their time sleeping, they do wake up and play. Yes, cat toys are important pet supplies. You can buy cat toys of every type, from tinkling balls and crinkling mice to feathery gloves and everything in between. If you don't want to spend big bucks on cat toys, take a look around your house for items to keep your purring pet entertained. Twist ties, pipe cleaners, rolled-up socks, ribbons, pieces of string and fuzzy pom-poms are all good choices to keep kitty happy and entertained.

- **First aid supplies** No pet owner wants to see a pet sick or injured, but it does happen. Be sure to plan ahead for that possibility. Put together a first aid kit for your new cat. Include any medications that your cat might need, such as de-wormer, flea repellent, hair ball treatment and bandages. For a more comprehensive list of what to put into a pet first aid kit, see "Pet Care Basics: First Aid Kit Essentials for Cats" on page 62.

- **Cat carrier** Most cats don't enjoy going for car rides. If you're taking them to visit Grandma or the veterinarian, you need to use a good cat carrier. This is one of the important supplies on your list. Close and confined spaces make cats feel more secure, and the cat carrier will provide that sense of security during a trip or an unexpected emergency. Cat carriers can be hard-side carriers or soft, collapsible carriers (which are space savers). Choose your favorite, then store it in an easy-to-access location in your home. If you choose a hard carrier, you'll need to line the bottom with a towel or soft blanket.

Last, but definitely not least, a new cat needs the love and attention of its new owner. Don't forget this important aspect of pet ownership. Cats may be somewhat antisocial, but they do enjoy affection and attention from the humans they know and love.

How to Introduce a New Cat into the Household

Introducing a new cat into your home requires more than a loving heart and an extra litter box. Your other pets may be put out by the addition of this interloper into the family. By following some of these tips, your odds for a successful introduction will increase, and the process of adding a new cat into your home will be smoother for you, the newcomer and the family pets.

 Evaluate the personality and gender of the other cat(s) in the home. If you already have a cat that is territorial and aggressive, introducing a new cat into your household may cause turf battles. Often de-sexed cats of a different gender and age are more able to become friends. Older cats can become resentful and stressed if a playful kitten is thrust into their territory and adopt bad behaviors, such as urinating outside the litter box.

 Evaluate the personality and breed traits of the dog(s) in the home. Certain dog breeds are not cat lovers, so check with your vet before you bring home a kitten to a dog that might see it as prey.

 Proceed slowly with the initial introduction. For a few days, avoid giving your new cat free range of your home. Temporarily isolate him in your bedroom or spare room to protect him from being hurt or frightened by your other pets. During this time, bond with him by visiting him frequently and speaking in low tones and in a calm manner. He will learn the scents of the other family pets and they will get accustomed to his odor during this time of separation.

 Allow the new cat to roam the house while your other pets are contained. Cats are curious and your new little one will be eager to test his boundaries. Let him gain confidence and get acquainted with the pet scents in other areas of your home by allowing him to be free for a few hours each day.

 Have controlled meetings. Begin a controlled introduction of all pets using one of two options. Either create a small crack in the door of the isolation room so that nose-to-nose sniffing can occur without any physical damage, or put your new cat in a travel case. This way you can move him around from room to room and he will begin to feel he's a part of the family without fear of attack from your existing animals.

 Let your pets be. Eventually you have to let your pets work out their relationships. They may have some disagreements and territorial spats, but usually both cats or kittens and puppies or dogs will find a way to coexist in the same home.

Normal feline behavior includes hissing, growling and swatting with a paw. Dogs will often whine, huff, bark or growl at the new pet. Rarely will a dog charge or bite the new cat, but it can happen, so monitor your pets until you are sure they are all getting along and you see no signs of aggression.

Does a Nonallergenic Cat Exist?

Almost 15 percent of the world's population suffers from some form of cat allergy. It is no surprise that some cat breeders advertise their breed as being hypoallergenic or nonallergenic, as this trait would be sought after by buyers who want to adopt a cat but can't because of allergies.

While it is true that there are breeds with a lower allergy-reaction rate than others, scientific research has produced no definitive results proving that a completely nonallergenic cat breed exists. Experts know what causes allergies to cats, but eliminating all the allergens seems impossible.

Understanding allergic reactions to cats

Most cat breeds possess high levels of the Felis domesticus allergen 1 (Fel d 1) protein generally found in cats' sebaceous glands but to a lesser extent also in saliva and urine. As cats groom themselves, the Fel d 1 protein is spread over them and their environment. The oft-maligned dander is merely the dead skin cells all creatures produce — but the dander does act as the carrier of the Fel d 1 protein.

People suffer from cat allergies when their immune system overreacts to the Fel d 1 protein, mistakenly seeing it as dangerous. Their body stores this information, and when they come in contact with the Fel d 1 protein again, the same false overreaction occurs all over again.

The company Allerca, Inc. claims that its biotechnology breeding program has produced a 100 percent hypoallergenic cat breed that lacks the Fel d 1 protein. As no independent studies have been done to support the claim, careful research to verify the validity should be undertaken before considering a purchase of this very expensive cat breed.

There are other breeders who also believe their cats produce lower allergic reactions. Some previously allergic owners may not suffer an attack around these breeds, while other people may consider even a low-level reaction better than a full-blown attack. These breeds share undercoat characteristics similar to the Siberian cat breed.

- Manx
- Ragdoll
- Turkish Angora
- Maine Coon
- Russian Blue

Before adopting a cat, allergy sufferers should spend time around a variety of low-allergy cat breeds before making a final decision on which breed to take home. Through experimentation with exposure to the breeds mentioned here, people with allergies may find a cat breed that, while not 100 percent nonallergenic, causes them the fewest and least violent of symptoms.

Best Cat Breeds for People with Allergies

Siberian: It is believed that Siberians produce less of the Fel d 1 allergen compared to other breeds. Although it has not been scientifically proven, many breeders advertise that the Siberian's nonallergenic qualities are due to a thick undercoat and oily top coat. This combination of coats keeps their skin hydrated and reduces dander production.

Sphynx: This breed is described as "hairless," but in fact the sphynx does have a very fine downlike hair that feels like suede to the touch. Like all cats, the sphynx does produce dander; the difference is that the effects of its dander can be easily controlled with simple body rubdowns.

Rex (Cornish, Devon, Selkirk, German and LaPerm): Many owners of the Rex breeds claim their cats cause fewer allergy attacks. Breeders back up this claim because they believe that the Rex's short, fine coat discourages the buildup of danger. Rex breeds also have no top coat and a very fine undercoating of hair, which some people say keep allergens to a minimum.

Playing with Kitty: The Best Cat Toys

An important part of cat ownership is understanding the importance of playing with your kitty. For kittens, playtime means learning and bonding with their owners. For older cats, playing prevents boredom and gives them a healthy workout to control obesity.

Cats are social creatures and will be happier and more content when playtime is a regular part of their day. When you play with your kitty with an interactive toy, the benefits are doubled as you both get exercise together. Yet when you are away, you still want to encourage playtime with toys that engage and interest your cat and keep her entertained.

Catnip mice, jingle bells and Ping-Pong balls are great toys for your kitty, but some toys bring more to playtime than just fun. The following toys exercise your cat's mind and body and have been rated by cat owners as some of the best cat toys you can buy.

- **Neko Flies and Mini Neko Flies** Most cats enjoy feathers, fur or catnip-stuffed toys attached by a string to the end of a stick. Because cats can turn into acrobats when playing with this type of interactive toy, they don't usually last very long. These Neko Flies are more durable than most, although many varieties can be found in pet stores. Flip them in the air or drag them slowly across the floor to bring out your cat's natural stalking skills.

- **Miracle Beam** Laser toys do not appeal to every cat, but when they do, the cat gets a good workout chasing the light beam across the floor and up the side of walls, and many will gaze entranced at the light dot on a ceiling. (Just be sure not to aim it into the cat's eyes.)

- **SmartyKat Bungee Bouncer Doorknob Dangler** Just hang the bungee, made with fur, feathers and fleece, on a doorknob and watch your kitty get a terrific leg and chest muscle workout.

- **Cat Ware Kit-E-Quiz Problem Solving Gameboard** A durable, plastic base that has a track with balls for kitty to push around. The fact that cats cannot get the balls out keeps them interested in the toy. While there are many versions of this toy found in pet stores, Kit-E-Quiz is unique because it is two-sided, making it two toys in one.

- **Seagrass Scratcher Post** Every home should have a kitty scratching post. Made from all-natural fibers, this post offers horizontal and vertical scratching surfaces. For added kitty interest, a string is attached at the top with feathers that dangle over the post.

- **2Gen Undercover Mouse** A floor base covered by circular fabric, it has an electronic arm with a mouse on the end that moves randomly, even reversing directions, under the cover. With the right speed control, you can spark your kitty's hunter instincts. The gearless drive provides product reliability.

- **Talk to Me Treatball** This cat toy combines entertainment and tasty treats for those times when you are away. After you record a spoken message especially for your cat, every time he moves the ball the sound is activated. When he hears your voice, a treat pops out. This toy can be helpful for cats that experience separation anxiety.

The Top 10 Most Popular Cat Names

From the color of their fur to their unique personalities, cats earn their names for a variety of reasons. All pet owners draw from their own favorite sources, whether that is current celebrities or sports heroes. To create a list of the 10 most popular pet names for cats, Bow Wow Meow tracked the purchases on its pet tag website for the monikers that cat lovers liked the best.

10 Patch If you bring home a frisky kitten that is a solid color except for one small colored patch, this name may immediately come to mind.

9 Simba In Swahili, this name means "lion," which is one reason Disney used it for one character in *The Lion King*. This name definitely would apply to a beautiful orange Maine coon cat, which can look a lot like a lion on the prowl.

8 Shadow One of those pet names that can do double duty, Shadow is a good choice for a jet-black cat, or one that disappears at all the wrong moments.

7 Sassy Some pet lovers say that their cat owns them, not the other way around. Sassy is appropriate for the feline that rules the roost.

6 Kitty Short for Katherine, a Greek name meaning "pure," Kitty also is one of those generic pet names applied to most cats.

5 Sam Short for either Samantha or Samuel, this name is the Hebrew word for "heard by God."

4 Smokey This is the ideal name for a cat that is one dark, solid color.

3 Max This name, which is short for Maxine or Maxwell, pops up near the top of many popular pet lists. That's appropriate, because in Latin, Max means "greatest."

2 Tiger Paying homage to its feline relatives, Tiger describes a cat that has beautiful stripes on its fur.

1 Tigger In the imaginary world created by A.A. Milne, the honey-loving bear named Winnie the Pooh had a bouncy friend named Tigger. Unique even among the animals in Milne's Hundred Acre Wood, this tiger-like creature had a rubbery head, a springy tail and a bouncy personality, just like many pet cats.

5 Online Pet Name Finders

Looking for the perfect name for your pet? There are thousands of possibilities. Fortunately, the World Wide Web might have the answer. The following are five pet name websites that can help you tap your inner creative genius.

1. **Bowwow.com.au** This site is a bit colorful and sometimes difficult to navigate, but it offers something that other sites don't: name meanings. In addition to searching its extensive database for the perfect name, you can also find out what each name means to help you find something suitable. You can click on the search field in the left navigation bar to search for pet names, male and female. You can even choose categories such as religious, foreign, sport names and nicknames. Once you have a list of possibilities, you can search each one to find out exactly what it means. The site will even tell you if any Beanie Babies share that name, or if there are any pop culture references.

2. **Funpetnames.com** If you are looking for a better organized website, this is probably right up your alley. Funpetnames.com is focused more on category: male or female, geographical region, famous names, type of animal and a host of other ways to search. Many of the names found here are also popular choices for people. Although this site lists the meanings of the pet name you choose, the descriptions are more limited and are related mostly to foreign translations. For example, the name "Ammitai" means "truth" in Hebrew. It also has lists of popular monikers throughout the ages.

3. **Petnamesworld.com** A little zany and a lot of fun, this site is a great resource for parents to peruse with their kids. Adorable little icons for each type of pet combined with wacky fonts and colorful backgrounds make this an enjoyable website to search. It is also informative, with more than 11,000 names in its database. The search engine is more limited, however, and you can only find names according to male or female animals, and alphabetically. You can keep a record of your favorite names for future reference, and it also has an extensive list of links to other informative pet-related websites.

4. **Mypetnames.com** Although pet names are certainly a focus for this website, there is a host of other information about breeds, wild animals and other cool facts. It won't win any design awards with its plain black, gray and white colors, but it is a bit easier to navigate than other sites and offers a good deal of information. This is the place to go if you're looking for a name for an exotic pet. You'll have fun browsing through all the ideas, which are ranked according to popularity.

5. **Pet-net.net** If you are looking for names related to a particular country or ethnicity, this is the website you will want to use. Browse for Celtic, French, German, American, Japanese and even Native American names in its extensive database. It also has a bookstore if you need a larger collection of possibilities.

All About Cats' Eyes, Tongues and Tails

You love every part of your cat, from the top of her head to the tip of her tail. But have you ever wondered what makes her so special? Why is it she can see so well at night? How come her tongue is so rough and scratchy? And what is she trying to say when she swishes her tail? Below are some of the answers to these common questions people have about their cat's eyes, tongue and tail.

Eyes It is true that a cat's night vision is much better than a human's night vision, but cats can't see in complete darkness as some folktales profess. Thanks to a special feature of the muscles in her eyes, your cat can see quite well with as little as one-sixth of the light that we need to see clearly. In the wild, cats perform much of their hunting at night, which makes good night vision a necessity.

Your cat has you beat when it comes to seeing in the dark, but a human's ability to detect color is usually far superior. Although cats are not entirely color-blind, their visual palette is very limited.

Despite superior vision overall, your cat cannot see directly beneath her nose. You may notice this when feeding your cat a treat. If you put the treat directly beneath her nose, she often has trouble finding it, even though she can clearly smell it.

Tongue Cats' rough, abrasive tongues have been compared to sandpaper. While you might enjoy your cat's kisses, too much kitty love and your skin is likely to feel a bit raw. If you look closely, you can often even see little barbs (called papillae) on the surface of your cat's tongue. These hooks are important since they help your cat groom herself, grasp food and hold on to struggling prey.

Cats are notoriously finicky eaters, especially in comparison to most dogs. However, this isn't a matter of superior class. Your cat actually has two sets of taste buds, which allow for a keener sense of taste. Cat tongues respond not only to flavor but also to texture as well, which is why dry cat foods come in such a wide variety of shapes.

Tail As most cat owners know, a cat's tail is a good indication of her mood. Unless your cat is asleep, her tail is nearly always moving. You may notice it bristle up and expand during a fight, twitch when she's irritated or swish from side to side when she's very angry or threatened.

Your cat's tail isn't just a barometer for her state of mind. It also serves the practical purpose of helping her stay balanced. Cats can walk across skinny banisters or fence rails with ease, thanks to the counterweight effect of their tail, even making sharp turns without losing their balance.

The next time you look into your cat's soulful eyes or feel the rough caress of her tongue or stroke her tail, you'll have a new appreciation for what these body parts can do.

Spaying and Neutering: For Our Pets and Feral Cats

While the decision of whether or not to spay or neuter your cat seems like a straightforward one, often pet owners don't realize the additional benefits that come with having cats spayed or neutered at an early age. More than just stopping unwanted litters of kittens, spaying or neutering your cat will help your cat live a longer, happier life as a contented companion pet.

Why it is important for our pet house cats to be spayed or neutered

Your cat will be healthier over its lifetime. Female cats can develop mammary-gland tumors, ovarian cancer and uterine cancer, especially if they go through repeated heat cycles without breeding. Male cats are at risk of testicular cancer and prostate disease. By having your cat spayed before she experiences that first cycle or neutered as soon as he is old enough, you can reduce or eliminate the risk of your pet developing cancers or other serious diseases.

Your cat will be happier over its lifetime. Cats that are neutered or spayed are not as likely to roam and be injured by cars or other animals. Females will not attract intact males looking to mate and will not be forced to bear numerous litters. Often "unwanted" male suitors can become aggressive while fighting for the attention of a female in heat, even biting and attacking the female if she resists. Intact males will also "mark" their territory, whether it is your house or car or white picket fence, by spraying it with their strong, foul-smelling urine.

You will be happier over your cat's lifetime. Spaying or neutering can extend your cat's life an average of three to five years or more. This gives you more time with your best friend and less time worrying about finding homes for unwanted or unexpected kittens, or making trips to the vet for injuries or diseases caused by your cat's drive to reproduce.

Why it is important for outdoor feral cats to be spayed or neutered

Population control Feral cat populations can quickly grow out of control, with females being able to begin breeding as young as four or five months and able to produce litters of up to eight kittens two or three times a year. It is possible for one female to have as many as 100 kittens in her productive years!

Disease control Feral cats are often wild, fighting with other cats or wild animals. Since they are seldom vaccinated, illnesses and diseases can become epidemic in a feral population. Since they defecate everywhere and not in litter boxes that are routinely cleaned, it is possible that uncontrolled diseases or parasites can be transmitted to house pets or people simply by walking through contaminated soil, which then is carried into your home on your shoes or feet.

Nuisance control Feral cats can become destructive in their search for food, mates or shelter. They can be aggressive and noisy and bring a strong, fetid odor to areas where they live. When the local animal control officer is called in to catch the feral cats, they are often too sick or too wild to be adopted and must be euthanized.

The Top 3 Most Popular Breeds: Maine Coon, Siamese and Ragdoll

Each year various magazines and Internet sites publish a list of the "most popular cat breeds" based on input from breeders, owners and veterinarians. There are three well-known breeds that consistently top every list. While these cats are distinctly different in appearance, they do share personality characteristics that make them so extraordinarily popular. The most endearing trait they share is their strong attachment and devotion to both their human and animal family members.

Maine coon These large-breed beauties with their gorgeous fur and extraordinary plumed tails can be found anywhere in the world because of their adaptability to a variety of climates. Most Maine coon owners would not characterize their cat as one that likes to be picked up, but would tell you he is their constant companion.

While the Maine coon may not choose to spend time curled up in your lap, you can be sure he is close by, watching your every move. Maine coons are known to follow their owners around the house, supervising each project the human is doing and often "lending a paw" to help.

A Maine coon cat loves to play and will interact well with children, dogs and other cats. Maine coons love to chase after small objects, which makes them excellent mousers. They also love water play, fetching, carrying small objects in their mouth and guarding their family members by sleeping by their bedroom doors.

Siamese This breed has long been the standard for elegance in the feline world. Owners of Siamese cats know that these cats are much more than just a pretty face. It is true that they will not hesitate to speak with their distinctive voice to remind you of their ancient, royal pedigree and your humble origins. But their imperial lineage and attitudes do not keep them from seeking your lap just to cuddle or needing your affectionate pats numerous times during the day.

While they have been accused of being temperamental, this label may have more to do with their inquisitive nature and their desire to be the center of attention. The beauty, grace, intelligence and sophistication of a Siamese cat are bonuses to a breed that is known for the abundant love and affection it gives to its human family.

Ragdoll Their funny name is derived from the way their body tends to "go limp" when they are picked up. Ragdolls are often called "dog cats" because they are the one breed of cat that shares the most characteristics with dogs. Although not a traditional "lap cat," they crave friendship and will sit as close to you as possible.

Whether relaxing or working, you can count on your Ragdoll "dogging" you. If you leave the room, your Ragdoll will get up too and follow you. Ragdolls are loyal and will stay by your side if you are sick.

This very loving, passive breed has a natural tendency to keep its claws retracted during play with its human family. Sociable and gentle, Ragdolls interact well with children and other animals in the home.

What Are the Best Cat Breeds for Lovers of the Exotic?

Most people are very content adopting a common house cat from the local animal control center, but some cat lovers seek out unusual or rare breeds when they make their adoption choices.

Domesticated exotic cat breeds usually are developed from a single cat possessing an anomaly such as a stump tail or no fur. The cat is then bred to retain the odd gene, passing it along to the kittens. Other times a domesticated cat is bred with a wild exotic cat. In the latter case, the breeder hopes to reproduce the wild cat's markings while conserving the domesticated cat's personality. These cats are sometimes referred to as hybrids.

Savannah The Savannah is the result of breeding a domesticated cat (often a Bengal) with the African serval, creating a striking, spotted hybrid feline. The Savannah possesses a normal house cat personality, but retains its wild ancestral appearance. The classic markings of a Savannah are a mixture of spots and stripes on fur colors ranging from silver to amber. The outside of the eyes has distinctive tear-stained markings. The Savannah can grow to be a large cat, often weighing 18 to 29 pounds at maturity. A low-maintenance cat, the Savannah needs no special food or care, but do check your local city, county and state laws to be sure owning a Savannah is legal where you live.

Korat The Thai people love Korats and consider this silver-blue exotic breed to be good luck. The Korat, also known as a Si-Sawat, is believed to have originated back in the ancient times of Siam. This rare exotic breed has a coat of silver-blue fur that has an extraordinary shine. Yet the most outstanding characteristics of a Korat are its large, round, expressive eyes; they start out blue, change to amber and end up a luminous green. This breed is known to be extremely intelligent, sensitive and gentle, with heightened abilities of smell, sight and hearing. Korats are small exotics with heart-shaped faces and rarely weigh more than 10 pounds as an adult. Breeders say the true beauty of a Korat doesn't appear until he is about 5 years old and has completed maturing.

Sphynx This hairless breed is perhaps the most exotic cat of all. The origin of the sphynx breed is claimed by more than one person, but we do know that the first hairless kitten was just a fluke of nature. To some people, these naked kitties are unattractive, but nearly everyone admits they have the most unusual appearance of any breed. Born looking like large-eared rodents rather than kittens, this breed grows into their ears and wrinkled, hairless skin. Although referred to as hairless, the sphynx does have a fine peach fuzz on his coat. Look closely at his skin and you can see the markings and colors he might have had in his fur. A sphynx is known to be loving, intelligent, curious and independent. Because a sphynx has no fur, be alert to the possibility of sunburn and keep him from getting cold or overheated.

What Is a Polydactyl Cat?

A polydactyl cat is literally "many-toed." The normal five toes on each front paw may be supplemented with one, two or three extra toes. The toes are either added to the outside of the paw, giving a snowshoe or hamburger-patty look, or located on the inside, forming a large thumb-like appendage. These felines are also known by the more familiar terms of "mitten cat," "boxer," "thumb cat," "six-fingered cat" and "double-pawed cat."

Polydactyly in cats is a genetic variation and not a deformity. It rarely impacts the cat's health. It is most often seen on the front feet and only occasionally on both the front and hind feet. True hind-feet-only polydactyly is thought to be genetically impossible. Hind feet almost always have the "patty" form of multiple toes as opposed to the "mitten" form.

Polydactyl cats are somewhat rare in continental Europe. One theory is that they were wiped out in the Middle Ages as probable candidates for witches' familiars. On the other hand, sailors considered their presence lucky on ships; they thought that broad-pawed cats would have better dexterity, making them superior sailors and mousers. To this day, a higher incidence of many-toed cats is found in colonial-era port cities such as in southwest England and the U.S. Eastern Seaboard.

Polydactyly in cats is thought to be a dominant gene and occurs equally in males and females. It is also found to a lesser degree in the big cats.

TIP: Polydactyl cats' nails often need weekly trimming. The "extra" nails can be askew, which can prevent the cat from wearing them down naturally.

CAT BEHAVIOR

No one really knows what's going on between a cat's pointy little ears, but we do know that cats are a lot smarter than people once thought. Cats communicate in a variety of ways: meowing, purring, tail wagging and rubbing against their human buddies. Cats have a good memory and react negatively to situations or events that frightened them in the past.

Cats learn quickly if it is to their benefit. But unlike dogs, cats are not genetically driven to please their owners. They adore us—but always on their terms. Each cat has a unique personality and each breed has a personality of its own too.

You can bring out the best in your cat by learning all you can about how cats think and act. By understanding why your cat is doing something you don't like, you can often find a simple and effective way to stop him.

The ABCs of Cat Behavior

No two cats are alike. Each cat has a unique personality and possesses both good and bad behavior traits. Some traits, however, are universal and can be identified in almost every single cat.

Active in the morning Your cat will do almost anything to get your attention and get you awake. If he's hungry and needs to be fed, or if he's an outdoor cat that needs to relieve himself outside, he will keep pestering you until you give up and get up. He will jump onto your bed and purr loudly, paw at your hand or knock items off shelves or dressers. He will chew on your fingers, nip at your nose or play with your feet until you cry uncle and start your day.

Active at night It is normal behavior for cats to "party" all night long. Females in particular will be up scouting the house for prey and will pounce on ants, spiders or your twitching toes hanging off the bed. Cats have supersensitive hearing, and in the quiet night hours they may hear a mouse in the attic or a cat friend strolling through the neighborhood, which is part of the reason why nighttime is a happy time for a cat.

The cat's meow Some cats, like those of the Maine coon breed, often have a supersoft meow that is almost inaudible at times, but all cats meow since it is a way to communicate with both people and animals. Cats will meow to get your attention, to express happiness, hunger or fears, and to ward off predators. Often an older cat will howl loudly in the middle of the night because of growing senility and loss of hearing. Loud meowing can also be a sign of illness.

The itch to scratch Clawing and scratching are perfectly normal behaviors for cats. A cat needs to sharpen his claws and remove the old nail sheath, allowing the new one to emerge. A cat's claws are important to his survival. Cats need them to catch prey and to defend themselves against other animals. Provide your cat with scratching posts so that his natural instinct of clawing can be exercised on a scratching post rather than on your furniture.

Biscuit making Most cats "knead" or "make biscuits" on soft surfaces when they are happy or sleepy. This back-and-forth action of their paws is a leftover behavior from when they were nursing kittens. Other universal signs of a happy cat are eye squinting, purring and drooling.

Dig a hole and bury it Cats defecate by digging a hole with their paws, relieving themselves and then covering their feces with litter or dirt. They urinate in a similar way. Outside cats find special spots that they like to use because there's ample dirt for covering purposes. Inside cats that use litter boxes need enough sand or litter to ensure they can bury their wastes too. Keep a clean, well-filled box or otherwise your cat might seek another place to go that suits him better.

10 Things to Know About Caring for Kittens

The time to prepare for the addition of a new kitten into your family is before she actually arrives. Read the tips below before bringing kitty home, and you'll be prepared to best care for her during her early months.

1 Quarantine It's important to keep the new kitten isolated from other family pets until a thorough veterinary exam has taken place.

2 Home base A new kitten should not be allowed to roam the house even if she is the only pet. Close off a small area that includes her litter box, bed, toys, food and water bowls. Place her litter box away from her sleeping and eating areas. After a few weeks, you can let her venture forth into your home if she's feeling confident and curious.

3 Kitten-proofing Remove all small objects, strings, electrical cords, breakable items and poisons from the areas the kitten can access. Kittens can easily chew and swallow items that can cause them harm. Cords and strings can wrap around their necks, cutting off their air supply.

4 Supplies Purchase the kitten's accessories before you bring her into your home. The supply list should include a litter box, litter, scooper, kitten food, bowls, bedding, toys, carrier and scratching post.

5 Litter training When you bring your new kitten home, the first place you should take her is to her litter box. Select a box with sides low enough for a kitten's short legs to enter and exit, and use a simple clay litter. A young kitten can be affected by strong-smelling or clumping litters and develop allergies or eliminate outside the box.

6 Feeding If you give your kitten wet food, you will avoid urinary problems in males and give your kitten a more nutritious diet than most dry foods can offer. Most veterinarians allow some kitten chow but don't recommend free-feeding of dry or dry only.

7 Health check Your veterinarian will examine your kitten to make sure she is healthy and has all the vaccinations she needs. However, you must continue with scheduled appointments for fecal exams, vaccinations and checkups, especially in her early months.

8 Cleanliness The kitten's feeding and litter box area should be kept clean at all times. Wash water and food bowls daily. Scoop the litter box every day and completely clean it out once a week. Keep the floor clean of any food or litter spills.

9 Grooming Begin to get her used to future grooming procedures like teeth brushing, nail clipping and coat brushing by gently rubbing her gums, holding her paws and stroking her fur a few times a week.

10 Companionship Some kittens, by personality or breed, can be extremely timid during their first weeks in your home. If your kitten appears frightened of you when you enter her space, just sit on the floor and let her come to you. Sometimes, a soft, low humming sound will remind her of her mother's purring and help her forge a quicker bond with you.

10 Things to Know About Caring for Mature Cats

Your cat is getting on in years and you may have noticed your one-time ball of energy slowing down, getting gray around the whiskers or maybe missing the litter box or refusing his food. Here are 10 important things to know about cats that are in their senior years.

1 **Mature cats make great companions.** Older cats are usually calm, quiet and need less supervision. He may just want to hang out with you now, rather than run around the house knocking over lamps while leaping at errant moths. By this age he knows the household rules and may be perfectly content to stay close to you on the sofa or in your bed. Age often brings out your cat's ability to be affectionate and loving.

2 **Older cats' sense of smell deteriorates with age.** Your kitty may not be able to smell his food anymore. Buy food with stronger odors, such as fish-based cat food, to spark his appetite if you notice he is not eating much. Have his teeth checked to make sure that dental problems are not keeping him from enjoying his meals.

3 **Mature cats need extra roughage.** Check out the consistency of your cat's feces in the litter box to make sure he does not need a stool softener or a change of diet.

4 **You may have to brush and groom your kitty more often as he ages.** His body will be less supple than it once was, and he will have a harder time reaching his entire body for cleaning and grooming purposes.

5 **You may have to help him wash.** If kitty has a case of "stinky butt," you may have to help him out by washing it for him with warm water on a soft, disposable rag. He may not have the agility at his age to reach his nether regions.

6 **Make your kitty comfortable in his old age.** Since your older kitty is losing fat deposits and becoming bonier as he ages, he should have a soft cat bed to sleep in. A box filled with blankets or a basket lined with one of your old sweaters will do just fine.

7 **Mature cats drink more water as they age.** Make certain to keep his water bowl filled with fresh water. Although offering water is always important to a cat's health, it is more important as your cat ages.

8 **Older cats may become forgetful or frightened.** They may call out for you in the middle of the night or the early morning. You need to reassure him that you are still there and that he is OK. If possible, keep your old boy in your room at night so he doesn't get confused.

9 **Mature cats start losing their hearing and their sight.** Now is not the time to relocate his feeding bowls or litter box.

10 **Don't let your older cat go outside alone.** If your cat is an outdoor cat, you will notice that with age he will stay closer to home and eventually choose to stay in for comfort and safety.

The Finicky Feline

In an ad campaign created to sell cat food, Morris the Cat became famous for being a finicky feline with a fussy appetite. But is it really true that cats are finicky by nature? Let's take a look at reasons why cats may appear persnickety when actually they are just attempting to communicate their wishes.

Food Cats can be finicky eaters at times. You offer wet food, then you offer dry food, but neither meal seems to appeal to your cat. Often your cat's lack of appetite can be traced to being bored with eating the same food every day. If your cat becomes a finicky eater, try serving him something different to eat, another brand of cat food or a different cat food flavor. Sprinkle some cat treats on top of his food or mix in a teaspoon of quality scraps, such as lean meat or cooked eggs. Avoid using plastic cat dishes, because plastics contain an odor that can be off-putting to your cat.

Water While a dog will drink out of a toilet bowl or dirty mud puddle, a cat may sometimes sit and stare at his bowl of water until you have to walk over to see what is so interesting. If the water dish is sporting "floaters" or "dog-mouth slime," you can be sure your cat won't drink the water no matter how thirsty he may be. Many cats choose to drink from a glass rather than a bowl, and the water must be close to room temperature to please them. Bottled water is their ultimate preference.

Finicky? Not really. Actually cats have sensitive noses and refined tastes, and dislike water that is saturated with chemical smells. No plastic dishes for water either, for the same reason; it's wise to choose stainless steel or china for pet food dishes. Change the water in your cat's bowl daily to keep the water fresh and free of crumbs, and your fussy feline will be more apt to drink water, which is important for his health.

Toys So you went out and spent $50 on toys for your new kitten, but all he wants to play with is the bread bag tie, your mascara tube and the dog's tail. A cat's sense of smell is almost as powerful as a dog's, and those new cat toys may still carry odors from the manufacturing plant. Chemical smells are turnoffs to cats. Crumbled paper balls, homemade sock toys and items that roll, like pens, pencils and lipgloss tubes, are much more fun than a plastic-smelling toy with a high price tag.

People Sometimes considered aloof or independent, cats can appear to be ignoring you when they are actually watching you closely to see whether or not you are a threat. Cats spend time evaluating your scent to get to know you and your mood. When you have no fear or apparent interest in them, they return the favor and ignore you.

Finicky cats do exist, but many behaviors labeled as finicky are simply outward expressions of a cat's natural genetic traits.

The Scoop on Litter Training

Congratulations! You've brought your new kitten home and now you're thinking: "What should I do next?" The answer is: Introduce Kitty to his litter box! After kitten food, it's the second most important purchase you will make.

The gear for litter box training

To litter-train a cat, you need three things: a washable litter box, litter and a scoop. At first the litter box needs to be small enough for a tiny kitten to enter and exit by herself. Later on, you can step up to an adult-size box when your kitten needs more space due to her larger size. Litter boxes with covers are sometimes rejected by kittens and cats because they can trap odors, so avoid this style box and choose a simple litter pan for best training results.

Start your kitten off using basic clay litter and avoid highly scented litter or litter with clumping additives. Allergies and asthma can often be avoided if a kitten uses the simplest, most natural litter in her box while she is young.

You will also need a sturdy scoop with holes for lifting out your kitten's waste while allowing the clean litter to sift back into the box. Pick a scoop that can handle the job now when your cat is a baby but can also function as your kitten grows into a full-size cat.

Setting up the box

Wash the new litter box in warm, unscented soapy water to remove any odd smells, dust or particles left over from the manufacturing process. Rinse and dry it thoroughly. Place about two inches of litter into the box, and put the box in an easily accessible, visible area of the home.

Potted plants should be located in an area away from the litter box. It is easy to get confused where to wee when you're a wee one.

After the big welcome

Start house-training by introducing your new kitten to her nice, clean box. Give a little scratch in the litter and your kitten is likely to get the idea immediately. If not, try placing your kitten in the box and help her scratch around by gently moving her paws in the litter. Digging a hole to bury waste is natural to a cat, and the kitten should get the idea that this is the place very quickly.

Litter box maintenance and care

A cat's sensitive nose can become offended by a stinky litter box, and your kitten might decide it would be better to go elsewhere to do her business, like your bathtub, the laundry basket or maybe that potted palm in the dining room. So scoop the litter box daily and wash it in warm, soapy water weekly or as needed. A tidy litter box makes for a contented kitten, a happy owner and a sweet-smelling home.

Provide a safe, quiet, private place for your cat's box and most kittens will start using it right away. Patience, kindness and a watchful eye on your kitten's house-training progress will pay off by giving you a kitty that will neatly and discreetly take care of her bathroom needs.

5 Benefits of All-Natural Pet Supplies

We are constantly bombarded with advice from holistic nutritionists and health practitioners to incorporate more natural products into our lives. But what about our cats and dogs? Perhaps it's time to look at natural pet supplies for our favorite companions.

What are natural pet supplies?

"Natural" can mean almost anything the manufacturer wants it to mean, because there are no regulations specifying what is or isn't natural. So be careful when reading labels for pet supplies you buy. Generally, natural means food, toys, supplements, bedding, etc. that are not made from synthetic ingredients or materials.

The following are some of the benefits you may derive from buying all-natural pet supplies.

1 **Your pet's health may improve.** Natural pet foods are more likely to provide your cat with better nutrition, because your pet gets fewer chemicals and preservatives and more quality nutrients and vitamins. Many cats that suffer from skin allergies often get better when switched from regular commercial pet foods to organic or natural pet foods. Because natural pet foods are made from leaner meats and whole grains (instead of simple, more sugary carbohydrates), your pet may be less likely to become obese. Obesity can lead to complications such as diabetes and arthritis.

2 **You could save money with some products.** Many expensive cat litters are made with baking soda to reduce the odor of urine. This litter generally goes for over $8 for a 14-pound box. But if you line the litter box with your own baking soda, you can buy plain litter for half the price.

3 **Natural products may be safer for your children.** Because natural products don't contain harsh or toxic chemicals, they may be less harmful if your child accidentally gets into them.

4 **You'll help protect the environment.** Most natural products are made with the environment in mind, whether it's the product itself or the container it comes in. Often, natural products break down and are absorbed more easily into the environment than the chemical agents that are used to make up less natural products.

5 **Your pet may be more comfortable wearing natural collars and leashes.** The standard nylon collar is made to last, not to give your cat comfortable protection. Try a more natural fiber instead, such as hemp webbing, which is durable but comfortable because it softens over time.

It's important to remember that not all natural products are perfect, just as not all less-natural products are bad.

Pet Mess Cleanup Guide

There's no question that our furry family members bring us oodles of joy, but every pet owner knows they can also leave a few unwelcome surprises around the house. Then there's fur to contend with, and nasty critters like fleas who hijack a ride indoors. But living with pets doesn't have to mean living with messes.

Accidents Happen

Even the most well-trained pet will have an occasional accident. The good news is, most stains can be prevented if you act fast. Getting to it sooner rather than later is key.

If It's Still Wet...

1 **Blot it up.** Get up as much as possible so it doesn't begin to spread into the carpet backing and padding. Scrape up any solids with a blunt-edged object like a spoon. Then place a thick layer of paper towels (white only, so the print doesn't bleed) or an old towel on top of the area. Press or stand on the towels to help absorb the liquid.

2 **Resist the urge to scrub.** Carpet fibers are twisted together, and vigorous rubbing causes them to come apart. Instead, continue blotting until the area seems dry, then weigh down the towels with a phone book overnight to wick up any remaining fluid. If a spot remains once the area dries, try the steps below.

If It's Dry...

1 **Dampen the area lightly.** Wet the spot with plain water using a sprayer bottle, towel or handheld spotting machine (similar to a wet vac machine). Avoid flooding the area, which can cause the stain to penetrate the carpet. You want to mist the spot, not drown it.

2 **Extract the water.** Using the machine or paper towels, soak up the water; repeat the process of misting and extracting until the spot is gone. In most cases, plain water will do the job.

3 **Take it to the next level.** If there's an odor or the spot doesn't come up, try a pet enzyme cleaner, found at pet, grocery and home improvement stores. Don't grab just any cleanser in your house. Products such as disinfectants, hard-surface cleaners, detergents and powdered deodorizers can bleach the carpet or leave a residue that attracts even more soil over time. Also avoid using ammonia or homemade cleaners, which may get rid of the odor—but could actually attract your pet for a repeat performance.

4 **Know when to call a pro.** Since vomit may contain stomach acids, food dyes and bile, it's particularly difficult to remove. Give it a go, but if you can't get it out, call a professional for help. Go to *certifiedcleaners.org* for a pro near you.

Get Fur Under Control

1 **Brush your pet.** Comb or brush to remove excess fur. Ideally, long-haired pets should be brushed daily; all others should be brushed weekly.

2 **Keep your pet's coat healthy.** Good nutrition helps your pet's skin and coat remain healthy so he'll scratch less and shed less. Nutritional supplements may also help; ask your vet for recommendations.

3 **Vacuum floors and furniture regularly.** Carpet rakes, which run about $10 at home improvement or carpet stores, help lift pet hair that a vacuum might miss. Sticky rollers do a good job of removing fur from upholstery, and they're less work than upholstery brushes.

Stop Flea Invasions

If you've ever had fleas in your home, you know it's a real battle. A flea can live inside your home for up to a year. If your pet has them, vacuum furniture and floors, getting underneath everything; discard the vacuum bag. Afterward, set off insecticidal room foggers, following instructions carefully. You (and your cat) will need to leave for a few hours while the insecticide kills adult fleas, eggs and larvae.

    Since one of the life stages of the flea is protected by a hard covering, it's less susceptible to foggers. That means you'll need to repeat the entire process two weeks later to ensure you get every life stage. To prevent future problems, "protect your animal with topical or oral products. They're effective because they disrupt the flea's reproductive and neurological systems," says Bernadine D. Cruz, DVM, of Laguna Hills Animal Hospital in Laguna Woods, California. Skip homemade remedies, such as garlic collars or sprays; they simply don't work.

Prevent Odor Buildup

1 **Get a black light.** Caught a whiff of something but don't know where it's coming from? Get an inexpensive handheld black light (around $20 at home improvement stores). When you wave it over the carpet, the spot will light up.

2 **Primp your cat's bed.** You change your own sheets often, so what about your pet's sleeping area? Buy a pet bed with a washable cover and launder it frequently.

3 **Clean your cat's hair.** Ask your vet for shampoo recommendations (some may work better than others for your animal), and get a checkup to ensure your pet doesn't have a skin or other health condition that's causing an odor.

Pet-Proof Your House

Is your house pet-proof? Turns out many hazards are hidden in plain sight. Animal expert Warren Eckstein points out a few of the pitfalls.

Plants If your pet gets into your lilies, azaleas or philodendron, it can spell bad news. All are toxic to dogs and cats, so place potted ones out of reach and keep an eye on your pet when he's in the yard. For a complete list of dangerous plants, go to *thepetshow.com* and click on "Pet Tips."

Electrical cords Cats are attracted to wires because they can feel the electricity running through them—it's like chasing a mouse. Decoy wires with tubing to prevent chewing, and for added safety, unplug electronics when you're not using them.

How and Why Do Cats Purr?

Dogs ooze love and affection for their owners, while cats show affection in more subtle ways. When we cuddle, stroke or praise our cat and she begins to purr, we believe this is her way of expressing love for us. While this belief may be true, there are many additional things that cause a cat to purr. But first, how does she do it?

How cats purr

Your cat's larynx, or voice box, contains her vocal cords and laryngeal muscles. Most experts believe that when a cat inhales and exhales repeatedly over a short span of time, the laryngeal muscles vibrate. This causes the glottis (the space between the vocal cords) to open and close quickly as the air passes through.

A cat meows only during exhalation, but purring is something she can do while inhaling or exhaling. As your cat's rate of inhaling and exhaling increases, her sound frequency may also rise and her purring may get louder. A cat's purr can range from a low rumble to a high-pitched trilling sound.

Veterinarians have noticed that if cats' laryngeal muscles become paralyzed, cats lose the ability to purr.

Although some cats may appear as if they don't or can't purr, most do. Some cats purr so softly that you need to place your finger under their chin to feel the purr. Purring is considered a mysterious but charming characteristic of both house cats and some big game cats.

Why cats purr

Cats typically purr when they are content, satisfied and feeling social.

Conversely, cats often purr when they are sick, stressed, injured or dying. Purring releases endorphins or "happiness hormones" in the brain as a way to self-heal or self-soothe. There is also evidence that a cat's purring can stimulate bone growth and bone healing and has other beneficial effects on an ailing cat.

Mother cats often purr during the birthing process. Whether they purr because they are contented, to soothe themselves, to soothe the kittens or all three reasons is unclear. Because newborn kittens have undeveloped eyesight, hearing and sense of smell, a mother's purring helps the newborns understand that mother is close by and encourages the babies to come and nurse.

Kittens often purr along with their mothers when nursing and will start purring immediately when given a whiff of their mother's scent. Kittens start purring as early as the first day or two after birth and continue purring throughout their lives.

Purring is your cat's nonverbal effort to communicate with you, to tell you she is happy or that she craves attention. A cat purrs when she is hungry and anticipating her meal. She will purr loudly in the morning if she wants you to get up and get her breakfast, using her loudest purr to wake you up and get you moving.

What's My Cat Trying to Tell Me?

Many pet owners readily admit that they talk to their cats, but fewer owners are willing to disclose that they get an answer back. Your feline friend may not be able to vocalize her wants and needs in a language that you can understand, but she certainly employs several other effective methods of communication.

If your cat wants to go outside or into your room, she may paw or scratch at the door until you heed her wishes. When feeding time rolls around, she may become more vocal or wind herself between your legs. If it's affection that she wants, she may bump and nudge her head up against your hand or face until she has your undivided attention. You'll know when she's scared if she hides under the bed, and you'll know when it's time to play if she presents you with her favorite toy. Other behaviors, such as refusing to eat or drink, or urinating outside the litter box, can be signs of health problems. Whether she knows it or not, your cat is letting you know that it's time to see the vet.

These are just some of the ways that cats communicate with their owners. It may not be English, but it's certainly just as effective. Every cat has a unique personality that includes different methods of communication. Want to know what your cat is trying to tell you? Watch, listen and pay close attention, and she'll let you know in her own special way.

Kitty Personalities: The Social Cat, the Hunter, the Loner, the Mother, the Alpha Male, the Scaredy-Cat

Although a cat's personality is not always clearly defined, there are many common cat personality traits including the social cat, the hunter cat and the loner cat. Cats often exhibit characteristics of all three of these personalities as they grow from kittenhood to adulthood to old age. Your cat may have one of these cat personalities or maybe a combination of them.

The social cat In the early weeks of a kitten's life, the mother cat helps her kittens become social animals by her example. If mother cat shows no fear of other pets, family members or other people, she teaches her babies that animals and people are safe to be near.

Mother might help teach baby how to be social, but other factors decide how social she will become. All you need to do is watch one litter of kittens grow up and you'll see that social ability differs from kitten to kitten.

You can recognize a social cat by her love for people and her desire to be near them. She is known to thrive on physical contact and appears interested in everything people do. Most social cats enjoy being held, petted, talked to and groomed. They will seek out your lap for a quick snooze. Social cats prefer homes that are active and have both children and other pets in residence.

The hunter Pampered cats, lazy cats, fat cats, actually all cats —if given the opportunity— will hunt. If your cat is an indoor cat, it will stalk the tiniest ladybug or housefly it can find. If your cat goes outside, it will go after bigger prey such as birds and mice. Hunting is part of the essence of a cat, and cats hunt for instinctual reasons as well as for food.

The exercise, both physical and mental, that is required to locate, stalk and kill prey is part of what keeps a cat physically fit and mentally sharp. Many indoor-only cats suffer from a lack of hunting and need to have play replace what they might do naturally outdoors.

The loner A loner cat tends to be aloof and may appear nervous or fearful. Maybe she was born this way or maybe a traumatic event caused her to withdraw. She doesn't like physical contact and may hiss, bite or scratch when approached. Loner cats typically spend much of their time in their hidey-holes or under the bed.

Some loner cats can have a relationship with a few people they trust. The relationship is usually driven by the cat's need to be fed rather than a need for attention or affection. Loner cats can become aggressive and territorial if they feel trapped or threatened.

All cats, especially loners, need tranquility, a regular routine and a safe place to eat, sleep and use the litter box. All cats should have places to escape to if they feel threatened, but for the loner it's a must for her mental well-being.

The mother This cat is a natural caregiver. She likes to nurture and protect the other cats or pets in the home and is sometimes overprotective in her role. She may wash the other pets, groom them, feed them and defend them even if they don't need to be defended.

If there are no other pets in the home, she may adopt a small stuffed toy, a felt mouse or even a sock as her baby replacement. These cats can also be protective of their humans and will hiss or growl at perceived dangers. The mother cat often puts her own safety or welfare second, always ensuring that her "children" are taken care of first.

Bring a new kitten into your multi-cat home, and you are bound to see a mother figure step forward, whether female or male. The mother cat personality is most often female, but male cats have also been observed exhibiting the traits of the mother cat out of necessity or inclination.

The alpha male Out of all the common kitty personalities, the alpha male can be the most difficult for the pet owner. The alpha male knows he's boss and he isn't about to let anyone forget it. Bring a new pet into the home, and sometimes the alpha male feels the need to stake claim on his territory. He might hiss or meow at the newcomer or even swat it a few times to establish dominance. Neutering helps keep the male cat from being overly aggressive, from roaming and from spraying to mark his territory.

The scaredy-cat This poor soul is afraid of his own tail. When a thunderstorm rolls in or a new guest visits your home, you know exactly where you'll find the scaredy-cat. Whether his favorite hiding spot is under your bed or in your closet, he'll be there cowering in the corner. Some kitties appear to be timid from birth, but others have developed into scaredy-cats because of abuse, neglect or other earlier traumatic experiences.

If you've adopted a scaredy-cat, do your best to make your home warm and welcoming. Let him have his hidey-hole, keep loud noises away from him and let him get accustomed to living in a safe home. Over time, and with lots of love, affection and patience, your scaredy-cat may come out of his shell. Your scaredy-cat may always be a bit timid, but once he learns that he can trust you, your "cowardly lion" can also turn into a fearless feline.

If you see one of these common personalities emerging in your pet, be sensitive to your cat's needs. Give the mother cat something to care for, neuter your alpha male so he can become a happy, healthy pet and provide a safe haven for the scaredy-cat.

Teaching Kitty to Scratch on Her Post

Cats instinctively love to scratch. They scratch to mark their territory, shed old nails, sharpen their claws and get in a good stretch in the bargain. But sometimes their favorite scratching post is the arm of your chenille sofa or the back of your favorite chair. Don't worry! You can teach your cat to scratch on her post and save your furniture from the junk pile. Here are some simple dos and don'ts that will help you get kitty scratching in the right spot.

- **Do** start training her when she's young. Kittens are much easier to train than older cats. The sooner you start teaching her to use her scratching post, the better.

- **Don't** put the scratching post near her litter box. Cats are very selective about what they do where. Also be sure to put your cat's scratching post in a location away from the sofa she's destroying.

- **Do** experiment to find the right post. Cats are as finicky about their scratching posts as they are about everything else. It might take some experimentation to find the one your cat loves. Happily, there's an endless variety available, so there's sure to be one your kitty will like.

- **Do** try a bit of catnip rubbed onto the post to encourage her to try scratching.

- **Don't** give her something that's not sturdy. Cats are smart and they don't trust things that wobble. If the scratching post isn't rock-solid, she's not going to give it a second scratch.

- **Do** get a post that will allow your cat to stretch to her full length. Cats love to stretch and using a scratching post is a great opportunity. Make sure your scratching post is tall enough to accommodate your cat's size.

- **Don't** force her. Never grab your cat's paws and push them onto the post. She doesn't like to be manhandled and won't associate scratching with anything positive. Coax her, but let her do it in her own time. Show her by kneeling down and scratching it with your own fingers.

- **Do** reward her for good behavior. When your cat uses the post, be sure to praise her. A little treat wouldn't hurt either. Positive reinforcement is your friend.

- **Don't** yell or punish her for not understanding the role of her scratching post. Your cat won't understand why you're angry. She'll learn to associate the scratching post with something bad and that undermines your goal.

- **Do** put double-stick tape on the arm of your sofa or the back of a chair or any other areas you are trying to discourage her from scratching upon. Cats hate the sticky feeling on their paws and will quickly learn where not to scratch.

- **Don't** give up. Teaching your cat to use her scratching post can take patience. Some cats will take to it easily, but with others you'll have to give them time. Be consistent and supportive, and before you know it, she'll be happily scratching away on her post and your sofa will live to see another day.

Cat Discipline: Is There Such a Thing?

Have you ever tried to discipline a cat? It doesn't work very well, does it? Cats love you very much, but they don't have a need to please you the way most dogs do. You pay for the cat food and give your furry feline a warm, loving home; shouldn't she respect at least some of the rules of your house? The answer is yes, but to get her to cooperate, you have to use discipline a cat understands without ever making her fearful and anxious. Once a cat is afraid of you, she will never respect you, never mind your belongings.

What not to do

No hitting, screaming, pushing or throwing things—ever. Cats have very good hearing, and yelling can hurt their ears. Throwing something at her will make her afraid of your home because she won't know why she was struck. Hitting or pushing is cruel since she's much smaller than you are and can be hurt by your actions. All these abusive methods will teach her is to be afraid of you, your hands and your home. She will learn to mistrust you without learning what it is that you want her to do or not do.

What to do

Watch a mother cat with her kittens or sibling cats having a disagreement and you'll learn how to "talk cat." You can use this language to discipline your naughty kitty.

Your cat may look at you as though you have grown a third eye when you start hissing at her. But you have to do something to get your cat's attention, and this will do the trick. Use hissing only when kitty is misbehaving and needs discipline. Curl your lip back and hiss at your cat, letting her know she is doing something bad.

Growling at your cat can also grab her attention. Growling means, "I'm angry!" If you want your cat to discontinue a bad action, this is one method of telling her to stop scratching furniture. Look intently at your cat and emit a low, guttural growl. This will certainly get her attention and stop her in her tracks from being naughty.

Most cats are naturally fearful of water. Don't spray your cat with a water-filled squirt bottle; flick a few drops of water at her instead. Dip your fingers in water, and flick the droplets at your cat's face. A few unexpected drops of water on her nose will let her know you are getting "spitting" mad.

Reward good behavior

Remember to reward your cat for being good. Praise her and she will understand from the pitch and tone of your voice that you are pleased with her and love her. If she's been using a houseplant as a toilet, but you then see her in her litter box, make sure to give her a treat and she'll associate using her box with a pleasant experience. If she's scratching on her post and not the furniture, treat her to a game of fetch or give her a pinch of catnip as incentive to use her post again.

How to Stop a Cat That "Sprays"

There are few disadvantages to owning a cat. Independent and easy to care for, cats basically take care of themselves. Fill their food and water bowls, empty the litter box, brush them and let them snuggle on your lap, and they're happy.

However, as many cat owners know too well, some male cats have the nasty habit of spraying urine. Not all male cats spray, and it is never done to spite you or to be naughty; he is simply marking his territory.

Here are tips to ensure that your cat never sprays or to stop him from spraying if he's started.

Stop the problem before it starts. The majority of male cats will never spray if they are neutered before 6 months of age. Spraying is an adult behavior exhibited most often by intact males.

Keep your cat indoors. An outdoor cat is more likely to spray after he's had an encounter with another male cat that has threatened his turf. Cats may begin spraying because they've witnessed the behavior of roaming males who spray bushes, tires and other objects. After the wandering male sprays a spot with his scent, your cat may feel the need to respray the spot with his own scent, creating a spraying war between your cat and the visiting male.

Invest in an antispray product. Many people have had great luck with a product called Feliway, which is designed to limit cat anxiety (and subsequent spraying). Apply this product on surfaces where your cat tends to spray.

Eliminate formerly sprayed objects. You can never be sure where your cat is going to spray next, but many cats do return to the same spot again and again. They may also have a tendency to spray on certain types of objects, such as paper bags or piles of dirty laundry.

Don't punish your cat. As smart as your cat surely is, he doesn't understand why you have a problem with his marking his territory. If your cat sprays in your home, it's natural for you to become upset, but any behavior that will increase his anxiety will incite him to spray.

Avoid sudden changes. Sudden changes to your home or your routine can also increase anxiety for your cat and lead to spraying. Avoid relocating your cat's food and litter box, and keep feeding and grooming schedules consistent.

Stay a one-cat family. Single-cat owners report less spraying than multi-cat owners. Even a new puppy can trigger spraying. Introduce any new pets gradually and monitor your cat's behavior for signs of stress.

Catch him in the act. If you see your male cat backing up to a wall, door, bush or any upright surface with his tail held straight up in the air, he might be preparing to spray. Make a sound like *pssst* and it might stop him from completing the act. Remember not to punish him for an instinctual behavior nor frighten him since this will exacerbate the spraying problem.

What's Up with Catnip?

Catnip is a special treat for cats, evidenced by their loud purring, cavorting and apparent "buzz" after sniffing or rolling in it. Manufacturers often include this cat-friendly herb in toy mice and scratching posts because most cats flip for catnip. But beware: Excessive amounts of catnip have been known to make kitty ill, causing nausea or diarrhea in some cats.

What is catnip, and why do cats like it so much?

Catnip, also called catmint, is a perennial herb of the mint family that grows wild in many areas, including North America and Canada. The leaves of the catnip plant contain an oil called nepetalactone that emits an enticing aroma to cats similar to a pheromone. Most cats react to this scent by becoming either euphoric, playful or sedated, but some cats can become aggressive after inhaling the catnip scent.

How much catnip should I give my cat?

Offer your cat less than a teaspoon of catnip at a time. Usually just a few pinches are enough to make your kitty ecstatically happy. The more catnip you give your cat, the less effect it will have on her mood and behavior.

Can catnip be harmful to my cat?

Despite the wide range of behavior cats may exhibit, from wild to groggy, there is no evidence that catnip given in recommended doses (less than a teaspoon) is harmful whether it is sniffed or eaten. The effects of catnip don't last long, and an excited, active kitty on a catnip high soon transitions into a relaxed kitty taking a catnip catnap.

What are the benefits of catnip?

Cats tend to be playful after enjoying a sniff of catnip, and even a lazy tabby will be inspired to run around and get some much-needed exercise. Small amounts of catnip have been known to aid cats with mild digestive problems and calm hyperactive cats.

Which type of catnip is most effective?

Fresh catnip contains more nepetalactone and is stronger than dried. You can grow your own catnip for a steady supply of fresh leaves. Start with just a few plants, because catnip, like most mint plants, spreads quickly. Catnip can be dried, refrigerated or frozen for longer effectiveness.

Should I buy loose catnip, catnip sprays or toys?

Try all three methods of introducing catnip to your kitty and see which method she prefers. Rub or spray a small amount of the herb on your kitty's toys, furniture or bedding, but don't add it to her food. Catnip inside toys can dry up quickly and lose strength, while catnip sprays can be helpful to introduce kitty to a new bed or scratching post.

Why doesn't my cat like catnip?

Although many cats love catnip and can't get enough of it, there are cats that aren't able to smell it or don't like the smell. Genetics can play a role in why some cats are catnip lovers and others aren't. Young kittens and older cats often show the least interest in the herb, while young, active cats seem most affected. If your kitty is a baby and not interested in catnip, wait a few months and try again.

How to Cope with the "Midnight Crazies"

It's 3 A.M. and Fluffy is patting your nose for attention. You hear a marble roll across the floor just as you get to the crux of a great dream. Something is nibbling at your toes as you roll over in bed. Then a loud meow echoes through your bedroom. You may be convinced your cat is trying to drive you batty with this disturbing behavior, but that's not the case.

Your cat is genetically prone to the "midnight crazies." It is in your cat's nature to be a perky pest past midnight. Many of your cat's frustrating behaviors relate to the nature of the beast: Cats are active at night because in the wild, their rodent prey and favorite meal are active and scurrying about in the dark.

To mitigate or eliminate your cat-related sleep disturbances, consider these suggestions.

- At night, confine your cat to an area away from your bedroom. Close your bedroom door, and it will be unlikely that cat play will wake you up.

- Catproof your home. Pick up the noisy toys such as bell balls or marbles. They prove irresistible to your kitty just before dawn.

- Tire that cat out! Give kitty plenty of attention and playtime during the morning and early evening. This is especially important for one-cat homes where owners work during the day.

- Keep food, water and a litter box accessible at night.

- Use gentle guidance. With a gentle flick of water or a noisy can of coins you can discourage unwanted night play and get the sleep you deserve.

Kneading and Other Baby Behaviors

The endearing kitten-like actions of kneading, suckling and even love-biting are leftover behaviors that originated in a cat's infancy and now remind kitty of the happy times he had with his mother and littermates. You'll most often see these actions when kitty is sleepy, contented and happy.

Kneading

Kneading is the rhythmic motion kitty makes with his paws when he's sitting on a soft substance. His claws flare out when he pushes into the material and then retract when he pulls his paws back. He will knead (or "make biscuits") on a blanket, a stuffed animal or even your lap, reliving the feeling of warmth and safety he had with his mother. Often the kneading behavior is accompanied by closed eyes, purring, and sometimes even suckling or drooling.

Kneading is a carryover from his time as a kitten. Kittens facilitate the flow of mother's milk by gently squeezing the area around the mother cat's nipples.

Kneading may also be a nesting instinct. Cats have been observed kneading while turning in a circle, pressing down blades of grass until they've created a hidden, cozy nest.

A cat's front paws contain scent glands under the base of the claws, so every time kneading occurs, so does scent marking. When your cat is kneading, sometimes painfully, on your lap, he is telling others that you belong to him!

This scent marking may explain why suckling kittens "claim" a specific nipple on the queen. If you move the kitten to another nipple, he will often reject that nipple in search of his favorite one that he's already marked with his own scent.

Suckling

Suckling is a less common behavior than kneading, although both behaviors can be expressed at the same time. Your cat may be happily kneading a blanket when he decides to suck on the blanket too. Some cats "nurse" on people's fingers, earlobes or noses. Other cats may suck their own foot or even their tail as a self-soothing action similar to human thumb-sucking.

In some cats this behavior fades as they mature, but others do it for their entire lives. Generally it is not necessary to discourage suckling; to protect your belongings, give him a soft blanket on which he can knead or suckle as much as he likes.

Suckling that becomes obsessive and results in fur removal or open sores does need to be stopped. Similar to people with obsessive-compulsive disorder, your pet might benefit from medication and behavior modification.

Love-biting

This kitten-like behavior is not an act of aggression but an expression of playfulness. He is nipping at you the same way he did his littermates. Your cat does not mean to hurt you but wants to inspire you to join him in some kitten fun such as grabbing, biting, stalking and pouncing.

Biting on your fingertips or your nose can also be a variation of suckling on these appendages. In the morning, your cat may nibble on your nose or fingers because he's hungry and has found this biting a sure way to get you up to make his breakfast.

Tips on Living with Multiple Cats

Cats are like people: No two are exactly alike. And kitty differences can cause sporadic problems for the multi-cat owner. Because of cats' individual personalities and proclivities, you can often run into unexpected problems with your kitty kids. Living with multiple cats can be very challenging, but the rewards are greater too. Here are some tips on managing the multitude.

 Good choices upfront save trouble later on. The combination of cat personalities in your home creates a dynamic that you have to learn to live with. Encourage the pairing of a relaxed older cat with a frisky kitten. The older cat provides a model and the kitten will energize the older cat. Pairing kittens and young cats of the same age is also positive. Their similar levels of energy will ensure that each cat always has an enthusiastic playmate.

 Provide ample space. Each cat will have a respite from the other, if they need one, providing you have enough hiding and resting places for your cat family. Often, extra space can keep conflicts at bay. Instead of fighting, even an alpha male will often retreat to a safe, dark hidey-hole when he's feeling grumpy or overwhelmed. When penned up together, cats, just like people, can easily get on each other's nerves.

 Provide personal food and water. Each cat should have two dishes that are his own for eating and drinking. Ideally, each cat should also have his own special bed. After a while, your cats may decide to share their food bowls or beds, but you can't predict that this sharing will ever happen, or when.

 You can't have too many litter boxes. It's best to have a litter box for each cat. Cats are very sensitive to smell, yet they have a great tolerance for their own special smell. They prefer to go in their own box because it carries their own personal odor. Cats are very tidy and will not use a dirty box. If their box is unappealing, they may seek out another cat's litter box, or worse, go somewhere inappropriate.

 Have playtime together. When cats can become jealous of one another, it is vital to give each cat an equal time for playing and petting. Play is very important to cats since they need to expend their abundant, natural energy. If a cat is tired and bored, she may seek release in negative ways, such as picking fights or scratching furniture.

When you have multiple cats, there is always someone to play with. Direct the play yourself and get in on the fun. It can be very entertaining to watch your kitty kids leap, run and chase, stalk a dangled string or fetch a wad of paper that you toss into the air.

 Be patient and relaxed in your expectations. While the dream of furry creatures sleeping in a pile on the sofa sounds irresistible, the reality is they may disagree on occasion. It may take time for friendships to be cemented. With your guidance, determination and adherence to these few good commonsense tips, your dream of one big happy cat family will come true.

Living with Both Cats and Dogs

He's an avid dog lover, but you've loved cats since you were a child. Can your cat and his dog live happily ever after? Of course they can. It is completely possible to have a happy home with cats and dogs living together, playing together and sharing their owners' time and attention without a whisper of a growl or a hiss to be heard.

- **Do be prepared to be patient.** Concerns over territory, food and affection can make your existing pets irritable or nervous around the new pet. For the best introduction, consider these helpful tips when "yours" and "mine" become "ours."

- **Start young.** Puppies and kittens that grow up together often don't realize that they are supposed to behave as enemies. As they play together, a trust between the two species can develop into a lasting friendship. Although adopting a puppy and a kitten at the same time might sound like insanity, this is often the best way to get your dog and cat to become best buddies.

- **Give them space.** Just as humans sometimes need their own space, so too will adult cats and dogs that you introduce into your home. Don't try to force a friendship by locking them into a room together and letting them fight it out. Always give them an escape route, whether it is a way out of the room or a way to get out of reach.

- **Ease into it.** Using baby gates to separate the two species keeps them apart yet still allows them to see and smell each other. This will help adult dogs and cats to become accustomed to each other without pressure or aggression.

- **Be patient.** Until you are confident there will be no aggression from either pet, continue to keep them separated, even if the process takes months.

- **Relax.** Pets, especially cats, can often sense tension from their owners. If you are stressed about your dogs and cats getting along, they will be too. If you can be relaxed around them, they will soon realize that there is nothing to fear.

- **Possession is nine-tenths of the law.** Always feed your dogs and cats separately and give your cats a private place for their litter boxes. The best way to avoid food fights or bathroom wars is to keep cat food dishes and litter boxes out of a dog's reach.

- **Respect differences.** Sometimes there are unknown factors that prevent a friendship from developing between your cats and dogs. It could be a previous trauma by an unrelated dog or cat or a trait of certain cat or dog breeds. In situations where it becomes apparent that they might never get along, the best thing you can do is to accept their differences and don't try to force them together.

- **Spread your love around.** The new fluffy kitten is so adorable, but old Bowser has been a family pet for more than 10 years. Remember to share your attention and affection equally with all your pets so jealousy never becomes a factor.

The Best Way to Introduce Your Cat to Your New Baby

If a stork is winging its way toward your home, most likely you have prepared for the new bundle by painting and decorating a nursery, buying toys, setting up a crib and purchasing diapers, baby food and clothing. The stork will be pleased with all the preparations and changes you've made to best care for the new little one, but what about your cat? Have you taken the time to prepare your cat for the new arrival too?

The best way to introduce your cat to your new baby starts months prior to the actual birth of your child. Because cats are creatures of habit, they do not appreciate change. Cats will benefit from getting accustomed to all the smells, sights and sounds associated with a baby before you bring your newborn into your home. Desensitizing your cat to the sensory stimuli of having an infant at home will help your cat adjust, adapt and accept the new baby with ease.

Tips for preparing your cat

- Play recordings of babies crying.
- Place baby powder or baby oil you plan to use with the baby on your skin.
- Invite a friend with a baby for a short visit.
- Turn on the infant swing so he can hear the sound.
- Pronounce your baby's name aloud and frequently.
- Keep the nursery door closed if you don't intend to keep it open later.
- Walk around your home with a doll in your arms.
- Move the litter box to the quietest room.
- Stop using the word *baby* in reference to your cat.
- Slowly scale back on the amount of attention you give your cat.

Introducing the cat to the baby

Once the baby is born, a family member should bring home a used blanket, undershirt or even a soiled diaper, and let kitty smell the baby's scents before the cat and baby meet. The cat should be praised and given treats after he's sniffed at the baby items to imprint a pleasant association with the baby's scent. Move the clothing from room to room so your cat realizes this smell is going to be in many places in the home.

On homecoming day, it is best to keep your cat in a quiet room to minimize stress on the new mother and the cat. When you are ready, let your cat have a brief introduction to the baby. Use a soothing tone of voice and give him a reward for his good behavior. If your cat seems anxious, stop the introduction and try again later. Never leave your baby and cat alone together, especially in those first few weeks when your cat might be confused, jealous or worried about the changes to his home.

If introduced slowly and correctly, your cat should quickly learn that good things happen when the baby is nearby. By eliminating stress on your cat, you'll help him adjust more quickly to the sights and sounds of the new family member, and kitty and baby may soon turn out to be best buddies.

CAT HEALTH

Cats are stoic animals that rarely show they aren't feeling well until they are very sick. By recognizing the signs that your cat is ill, you get a head start on healing your cat before the illness worsens. Educating yourself about potential health problems common to cats is another way to keep your cat well and strong. Preventing common ailments like urinary infections or constipation can keep your cat out of the vet's office and in fine health. Knowing what is poisonous to cats helps you rid your house and yard of potential toxins.

Diet, environment, exercise and genetics all contribute to a cat's general well-being. Poor-quality food, infestations of worms or fleas, and missing vaccinations can rob your cat of good health. Today many of our indoor cats are suffering diseases brought on from a combination of obesity and boredom, and recent studies show that a cat's dental health may very well predict how long the cat will live.

Pet Health 101: Kitten FAQs

Adopting a sweet kitten seems like the most natural thing in the world. But many people bring home a kitten and have very little idea of how to feed it or take care of its pet health needs. Here are some of the most important questions you should know the answers to before adopting a kitten.

How old should a kitten be before you take it from its mother?

A kitten should be 12 weeks old before leaving its mother. At around 10 to 12 weeks, a kitten becomes socialized by its siblings and its mother. They learn how to play without being hurtful to each other. A kitten taken away earlier than this may end up being aggressive because it hasn't learned to curb that aggression.

What are the most common pet health problems young kittens face?

Ringworm This is a fungal skin condition that is very contagious among cats and kittens in close quarters. Many people come home with a kitten that has ringworm, which can very easily spread to people, too. Symptoms include a red, scaly, itchy spot that develops a red halo around it, resembling a worm. There may also be hair loss in the area. Ringworm in kittens is treated by regular shampooing with a special soap and by applying an antifungal agent, both prescribed by your veterinarian. Treatment can be expensive because of repeated trips to the vet to make sure the ringworm has been totally eradicated. Before adopting your kitten, find out when the breeder or shelter last treated its feline population for this annoying skin condition.

- **Ear mites** If you notice your kitten shaking his head a lot or vigorously rubbing or scratching at his ears, chances are he's suffering from ear mites. These are microscopic bugs that live in your kitten's ears and can drive him crazy from the relentless itching. This is a common pet health problem that is easily treated with topical medications that are rubbed into your cat's ears. Left untreated, ear mites "can cause rupture of the eardrum and inflammation of the middle ear, resulting in balance and coordination problems," according to ManhattanCats.com.

- **Worms** A checkup for intestinal worms should be a routine part of your kitten's health care. There are several kinds of parasites—roundworms, hookworms and coccidia—that can thrive in your kitten's digestive tract, robbing her of valuable nutrients. One of the worst offenders is coccidia, which can cause severe diarrhea and even life-threatening dehydration in kittens. On your first visit to your veterinarian, bring a fecal sample for analysis. Fortunately, worms in cats are easily treated.

- **Fleas** Watching your kitten scratching and scratching can drive you crazy. Fortunately, there have been major breakthroughs in flea control, including oral and topical products that may also control other internal and external parasites, according to ManhattanCats.com. The problem with these products is that the kitten must be old enough and large enough to be able to take the medicine.

What are the most serious pet health problems that kittens face?

Feline leukemia (FeLV) and feline immunodeficiency virus (FIV, which is in the same retrovirus family as FeLV). The feline leukemia virus in particular is a death sentence for kittens that contract it. The American Association of Feline Practitioners (AAFP) and the Academy of Feline Medicine (AFM) recommend testing all kittens by eight weeks of age for this virus before you bring them home, especially if you have other cats. If the kitten tests positive, the test should be repeated and confirmed. Unfortunately, kittens who test positive live only a few weeks after the diagnosis is made.

If you want to get a second kitten, when is the best time to get one?

Right from the beginning. If you think you may eventually want two kittens, adopt two kittens at the same time, preferably siblings. This way they already know each other and will grow up together and keep each other company. If you already have an adult cat, you can bring in a kitten as long as your first cat is only a few years old. Younger cats are more able to adapt themselves to a newer member of the family. Cats older than 8 have a harder time because they are often pestered by an energetic and rambunctious kitten.

Remember, taking care of problems when your kitten's small gives it the best chance to grow into a healthy cat. So, when you adopt your kitten, let your veterinarian be the first person to meet your new family member.

Pet Care Basics: First-Aid Kit Essentials for Cats

Though a veterinary clinic always is the best place to take your feline friend in the event of an emergency, you should be prepared to act before getting there. Part of properly treating your pet following an injury is staying calm and taking control of the situation. For top-notch treatment, cat owners should create an emergency kit. Following are some first-aid pet care essentials you'll be thankful to have on hand in an emergency.

- **Antibiotic ointment** Prevent infection by buying a high-quality triple-antibiotic ointment to apply to your cat's wounds. Having this product in an emergency kit is essential to ensuring a small injury does not escalate into a more dangerous, infectious threat. Many veterinarians and pet groomers use styptic powder, an antiseptic clotting agent, to stop or slow the bleeding of minor cuts and open wounds sustained by cats.

- **Bandages and medical tape** Immediately upon getting injured, a cat can be wrapped in nonadhesive gauze pads to control bleeding and prevent infection. Keep several gauze rolls on hand so you are prepared to bandage your cat before it is transported to a veterinarian. In addition to being used to secure bandages and cuts, medical tape is helpful in creating splints and tourniquets.

- **Cat first-aid book** Have a book about first aid for cats on hand. In an emergency, if you can't get professional help over the phone, you'll want to have a reliable resource nearby— and you don't want to be at the mercy of a sporadic Internet connection. There are several comprehensive reference books available, such as *The First Aid Companion for Dogs & Cats*.

- **Emergency contact numbers** Compile a list of important contact numbers in case of a pet emergency. These should include phone numbers for the poison control center, the nearest animal hospital and your veterinarian.

- **Ice pack** If your cat sustains an injury that causes severe swelling, use an ice pack to help reduce the swelling.

- **Medical records** Keep a file of any medical records you receive from your veterinarian. Also, keep documentation of any pet insurance you carry. Having these things immediately available increases the probability that your cat will receive prompt, effective care.

- **Muzzle** A muzzle is vital to administering care to a cat that is prone to attacking, either because it feels threatened or because it is in pain. By affixing a muzzle, you protect yourself and your cat from unnecessary harm, helping to ensure your pet gets care quickly.

- **Nail clippers** Regular maintenance of your cat's claws is important to his or her hygiene. But trimming a cat's claws is very different from trimming human fingernails. You need to be careful not to cut near the quick, which is pink and contains blood vessels and nerve endings.

- **Pet carrier** You should have a pet carrier in an easily accessible location in your home. If you need to transport your pet to an animal hospital, he or she should be placed within the carrier for the journey—for both your safety and the safety of your cat.

- **Rectal thermometer** The healthy temperature range for cats is 100°F to 103°F.

- **Tweezers** Alleviate pain from thorns, splinters and other foreign objects that get embedded in a cat's paws or body with a pair of tweezers. Magnifying pet tweezers are ideal for those hard-to-find objects causing your cat pain.

Warning Signs Your Cat Is Sick

Your cat can't tell you when he's feeling sick, so it's up to you to recognize the subtle physical or behavioral changes that indicate he's not well. Because cats are good at hiding the fact they're ailing, you might miss a serious illness if you don't pay attention to his daily actions. Here's a quick guide to help you recognize some of the telltale signs that your cat is sick.

- **Poor appetite** Cats are famous for being finicky, and it's not uncommon for cats to skip a meal now and then. But if your cat doesn't eat for two days or drink for one day, he should be seen by his veterinarian.

- **Lethargy** Cats sleep an average of 13 to 16 hours a day, so it can be difficult to tell if your cat is sleeping more than average. If you notice that your cat seems to be sleeping more and is less energetic, enthusiastic or social, it can be a sign he's not feeling well. Hiding in closets or under the bed also are signs of a distressed cat.

- **Vomiting and diarrhea** Cats throw up easily. The occasional rejection of food, heaving hair balls and using grass to help regurgitate are normal. But if your cat throws up several times a day and refuses food, there's something wrong. If your cat's vomiting persists or you see any blood in the vomit, take your pet to the vet.

  Diarrhea can be dangerous if left untreated. The occasional loose stool isn't anything to worry about, but if diarrhea persists for more than a day, take your cat to the vet. Both vomiting and diarrhea can lead to dehydration, which very quickly becomes a serious danger for cats.

- **Changes in urination** If your cat is urinating more or less than usual, it's a sign that something's wrong. Increased urination can be a sign of liver or kidney disease and diabetes. A decrease in urination can be a sign of a urinary tract infection or blockage. For male cats especially, urine retention can escalate into a life-threatening illness in hours.

- **Coughing or wheezing** Coughs can be signs of anything from respiratory infections to heart disease. Wheezing and other breathing issues can be signs of asthma. If your cat persistently or repeatedly coughs, take him to the vet.

- **Skin and hair problems** In many ways, a cat's coat is a reflection of how the cat is feeling underneath all that fur. If your cat's fur is oily, dry or looks patchy, it's a sign he's ailing. Hair loss and dry or itchy skin can be caused by fleas, ticks or mites. Dull, dry coats can be caused by poor nutrition and infections. Oily coat can be caused by the overactivity of the sebaceous glands predisposing to feline acne which is seen in varying degrees of severity.

- **Growling or crying when touched** Cats that have broken bones, were poisoned or are brewing infections from bites or other injuries are in a lot of pain and may growl or cry when touched. Be careful how you handle your cat so as not to hurt him or yourself, and get him to the vet immediately for emergency care.

Feline Health Facts

Cat owners are guilty of some of the same basic mistakes as dog owners—plus one error in particular that's specific to felines.

Fat Cats

Felines are twice as likely as canines to be obese, according to the Association for Pet Obesity Prevention. Most people know to walk their dogs, but they don't think of exercising their cats. If you've got a tubby tabby, tease him with a laser pointer or motion-sensitive toy to get him moving. Also, follow package feeding instructions for food amounts, and say no to table scraps.

Kitty Coverage

Cats are predisposed to their fair share of illnesses, from certain cancers to infection-related problems, so they need health insurance or a "kitty fund."

Rules, What Rules?

Think cats will follow your commands? Good luck! Just because your kitty scurries off the kitchen counter when you yell *Bad!* doesn't mean you're training her. All you're really teaching her is not to do it in your presence. Cats are more likely to train us than we are to train them. So what can you do? As veteran cat owners know, you simply learn to live with it.

Litter Box Faux Pas

If your cat has left you a surprise or two behind the couch, a dirty litter box is probably to blame. Cats crave cleanliness, so don't leave messes in the box more than a day, and scrub it twice a year with a 10% bleach solution. A dirty litter box can actually be dangerous. Cats will only use a filthy box once a day, and the more they hold it in, the greater their odds of developing a life-threatening urinary blockage. Obese, sedentary cats are at special risk. Since cats are territorial, if you have more than one, be sure to put out a box for each.

If you're tempted to ward off odors with air fresheners, liquid potpourri (which is toxic if licked by pets) and heavily perfumed litters, don't. Cats dislike strong odors and are sensitive to essential oils. Instead, use baking soda–scented litter and always fill the box two-thirds full to keep urine from soaking into exposed plastic.

All-Natural Treatments for Minor Pet Health Ailments

Pets can be rambunctious and can get into all kinds of pickles. From bee stings to minor scratches to diarrhea to hair balls, it may seem like your pet invites trouble. You care about your pet's health, but you don't want to act like an overanxious mommy by calling the vet every time minor ailments arise. Some education on all-natural home remedies for proper pet health is in order.

- **Treat bee stings.** After removing a bee stinger from your pet, you should prevent swelling and dull the pain. Mix 1 Tbsp baking soda with water to create a thick paste. Apply this to the infected area. You also can apply a cold ice pack to the injury.

- **Soothe mildly burned skin.** Carefully clip most of the fur from the burned area, cleanse it with mild soap and water, and apply aloe vera to the wound. Aloe vera will soothe your pet's skin, so don't hesitate to reapply it several times throughout the day.

- **Stop diarrhea.** Pet owners hail puréed pumpkin as a great all-natural treatment for diarrhea. Your dog or cat will love the taste of this natural healing agent for maintaining pet health.

- **Remove hair balls.** For most animals, hair balls are a mere annoyance. But there are several all-natural home remedies that unclog hair balls from an animal's stomach. Give your pet some butter to lick, and the hair ball will pass easily. Similarly, if you stick petroleum jelly on your pet's nose, it can lick off the slick substance, which will help move hair balls from its stomach.

- **Heal minor wounds.** Mixing apple cider vinegar with water will cure many small pet wounds/injuries. Simply dilute (1:1) solution of water and apple cider vinegar and apply it to your pet's wound with a clean cloth. For more serious wounds, use Dy's Liquid Bandage, which contains beeswax, herbs and olive oil.

- **Eliminate skunk odors.** When it comes to removing skunk odors, the long-discussed tomato juice remedy is, unfortunately, a myth. But there is a great all-natural treatment within your grasp. Mix water, 1 qt hydrogen peroxide, ¼ cup baking soda and 1 to 2 tsp liquid soap, and apply it to your pet's fur as you would shampoo. Rinse thoroughly after several minutes. If the smell doesn't go away, feel free to reapply this solution.

While minor pet injuries often can be treated with home remedies, you should never take pet health for granted. If your pet's injury is more serious than you can safely handle, do not hesitate to call the vet immediately. Your pet's health always should be your first priority!

Pet Care Basics: How to Choose a Veterinarian

If you're like most Americans, at some time in your life you've moved to a new town and had to choose a new family doctor. It's the same for your pets, except for one major difference: You're the one who decides if the veterinarian is a good match for your pet.

Don't take choosing a veterinarian lightly; your pet's health and well-being depend on your choice. And remember: The best pet care is preventive pet care, so once you've chosen a veterinarian, be sure to return for annual checkups.

Here are some tips on how to choose the best veterinarian for your cat or dog:

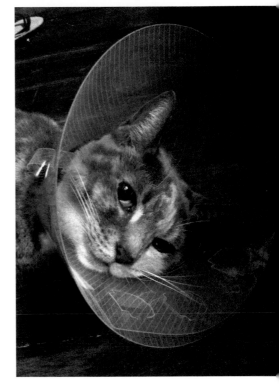

- **Start your search for a veterinarian before you need one.**
  It's better to invest the time and find a good vet before your pet gets sick, so you don't have to start looking for a vet when you're scared and under pressure.

- **Ask friends, coworkers or relatives to recommend a veterinarian.**
  Ask why they like that particular vet. Is he or she gentle with the animals? Affordable? Does the vet help you make decisions and give you options rather than telling you what you must do?

- **Check your town's Yellow Pages for vets nearby.**
  This can be a good place to start. There are also many online sources that can help you locate a veterinarian. One is LocalVets.com. Also call your state veterinary medical association for specialist referrals in your area.

- **Drop by the vet's office to make an appointment.**
  See how you feel in the waiting room. If the vet's office is really posh, consider that you'll be paying higher fees for your pet care than if the office is nice and simple. Also take into account how easy or difficult it will be to get to the veterinarian. Ask about the office's regular hours and emergency care after hours and on weekends and holidays.

- **See how you feel about the reception staff.**
  Are they willing to answer your questions? Do they seem to know what they're talking about? And most important, do they seem to like animals? If you have a good feeling about the office, make an appointment. It's fine to start with a brief checkup, just to establish a relationship with the veterinarian and to ask any questions about your pet's needs. When you meet the vet, make sure you and your pet feel comfortable. If not, keep looking. No one says you have to stay with the first one you meet.

Do You Need Pet Health Insurance?

Pet owners today who value their pet's health generally consider pet insurance a good idea. Because medicine has improved dramatically over the past several years, it is possible to provide a very high level of care for your pet. Because there are very expensive procedures performed on dogs and cats, pet insurance can provide peace of mind for pet owners. Pet insurance can also be the right and responsible move if you are bringing a new pet home.

- **Health coverage** Having a pet insurance plan will provide a range of coverage, depending on how much you pay into it per month. Having pet insurance provides assistance for large and unmanageable health bills that may occur. Many pets will need emergency care as a result of unforeseen circumstances, and preventive health care coverage and maintenance coverage can make having a health plan for your pet an easy decision.

- **Financing pet health treatments** If your pet has genetic conditions or has a chronic disease, having pet insurance will allow you to provide a high level of care for your pet while controlling expenses. This is much better than having to pay the entire cost upfront, and repeatedly, based on the number of office visits you need to make. By opting for pet insurance, you'll not only get to pay premiums in installments, but also much of the cost above the annual deductible will be reimbursed.

- **Insurance for wellness** You can purchase an additional wellness plan on top of your standard insurance to cover routine checkups, dental cleanings and exams, blood panels, microchip installation, spaying or neutering, and other common procedures. Because most pet insurance providers operate in a manner similar to a PPO, you will pay the costs of pet health upfront and then be reimbursed by the insurance provider. You will be happy with a wellness plan because it generally pays for itself.

- **Insurance for prescription medications** Prescription medications for cats and dogs can be very expensive to purchase without insurance. Having an insurance plan takes away this expense and makes it either free or extremely affordable, depending on the plan you purchase. Older animals may need prescription medications regularly to maintain pet health, and animals with genetic problems take medications often.

Pet insurance can be an essential part of providing for your pet's well-being. Ensure that you get the right plan for your pet by comparing plan services and deductibles before making a decision.

Why Petting Is Good for Your Cat—and for You

Touching is powerful medicine. It can bring hope, happiness, a feeling of well-being and a connection to others. It can help heal a broken body or a broken spirit. We all need to touch and be touched and so do our pets.

Petting is good for your cat

Just as we love it when our kitty rubs against our leg to show his affection for us, most cats like it when we give them attention too. Petting your cat adds to your cat's overall feeling of contentment and can also help reduce any stress she may be feeling.

As you stroke your cat, your hands will feel any lumps or scabs that might be on her body. Ticks can be felt and removed. Small wounds can be examined to see if veterinary care is necessary. Clumps forming in her fur can be more easily removed if she is petted daily.

If your cat is feeling anxious or ill but will still allow you to touch her, try petting, since it can be very therapeutic. When you pet a cat, she tends to purr. Purring helps your cat heal because the act of purring and the vibrations created during the process assist in keeping her calm and relaxed.

Massage therapy, which is an extension of petting, is now being used in some instances to help cats suffering from arthritis and other body aches and pains. Petting and massaging can help with circulation and stress reduction. When cats start to feel better, their brains release chemicals called endorphins that act as built-in pain relievers.

Petting is good for you

In addition to being companions and nonjudgmental confidantes, cats can be very beneficial to both our mental and physical well-being. Having access to a cat is almost like having a free doctor in the house. That beautiful furry little friend, so soft and cuddly to the touch, can help ward off feelings of depression, isolation and loneliness, especially for the elderly or people who live alone.

In addition to enhancing our quality of life and giving our mental health a boost, petting our kitty can also help reduce some of that stress and anxiety we have that can lead to headaches, high cholesterol, heart disease and many other physical ailments. Studies have shown that the simple act of petting a cat can lower your blood pressure.

When not to pet your cat

Although most cats love to be petted, some cats don't like it at all. Cats can actually get overstimulated from petting or even feel physical discomfort. Rubbing a cat's fur can cause static electricity, which can irritate the nerves in the cat's skin.

If your cat starts swishing her tail, showing her claws or other signs of irritability, stop petting immediately. Try again another time, petting a few strokes at a time starting at the back of her head and moving down her body. Don't try to pet your cat's tail, stomach or groin, since many cats dislike being petted in these areas.

70

Cat Nutrition: Canned Food or Dry?

Cats are carnivores. To be healthy, they need to consume a diet high in protein. Their diet should consist of meat, organs and fat with little or no carbohydrates. Cats are actually obligate carnivores, which means they cannot survive without meat.

Cats aren't able to digest grains and plants, so it is widely recommended that cats eat wet food as their main diet and have dry food only on rare occasions. Many cats, if they could talk, would say, "But I love dry food!" Additives in dry food often include sugar and starch that appeal to the cat but offer little or no health benefit.

Dry cat food

Before purchasing dry cat food, read the ingredients label carefully. Many brands, even premium brands, have a grain product in the first three ingredients, usually a form of corn or wheat.

Not only is a cat unable to receive any nutritional benefit from this filler, but research is beginning to show a direct correlation in cats between eating these grains and increased disease. Veterinarians have linked excess carbohydrates with feline diabetes, chronic renal failure, inflammatory bowel disease, cystitis and allergy-based chronic diarrhea.

Meat by-products should also be avoided. The term "by-products" is a euphemism for the inedible parts of animals, such as chicken feet and combs. If the ingredient is listed as "meat" (not *chicken, fish, lamb,* etc.), one must wonder just what form of "meat" the manufacturer is putting into its cat food.

There are some benefits to dry cat food: It spoils more slowly in the bowl than wet food and it can be left out for cats that graze throughout the day. Dry food is generally a less-expensive feeding option than canned food, too. If you choose to feed mostly dry food, it is crucial to keep a supply of fresh water available to your cat since dry food has little moisture.

Canned cat food

Canned cat food is closer to the natural foods a cat would eat in the wild. It is about 70 percent water, which is equal to the amount of water in cats' common prey. It is high in protein and low in carbs and contains the necessary fat to create a complete diet.

The high moisture content in canned food keeps your cat well-hydrated and promotes good health in general. It specifically helps male cats avoid urinary tract infections and blockages. Wet food helps cats stay "regular" and forces your cat to get some water in her diet even if she's not a drinker.

One of the only negatives of canned food is its expense. Quality wet foods are costly; wet food dries out if left in the bowl and must be thrown out.

Everything about a cat's anatomy indicates that it is ill-equipped to handle the plants and grain consumed by other animals: from its smaller digestive system (compared to the four stomachs in a ruminant that lives on grass and grains) to teeth that can only tear and not grind to the inability to obtain nutrients from grain. When choosing which food to feed your cat, it is important to keep these anatomical facts in mind.

Toxic Foods for Cats

You may have thought that feeding your cat "people food" is OK. It's true that some foods—such as lean cooked chicken, beef, lamb or scrambled eggs—are fine foods to feed your kitty in moderation.

But many foods that are healthy for humans are toxic or even deadly to cats. You can help maintain your cat's optimal health by making sure these top-10 poisonous foods are kept far away from your feline.

- **Mushrooms:** Can cause shock and even death to your cat.

- **Macadamia nuts:** Can adversely affect a cat's nervous and digestive systems.

- **Baby food:** Can contain onion powder, which can cause nutritional deficiencies.

- **Alcohol:** Can cause coma or death in a cat.

- **Grapes and raisins:** Can cause severe kidney damage.

- **Fat trimmings:** Can cause pancreatitis.

- **Citrus oil extracts:** Can cause vomiting and liver damage or failure.

- **Caffeine:** Can affect your cat's heart and is highly toxic.

- **Raw fish:** Can result in vitamin deficiency and lead to seizures and even death.

- **Onions:** Can destroy red blood cells and cause anemia. Garlic is slightly less toxic.

WARNING: Contact your veterinarian immediately if you suspect that your cat has ingested any of these toxic foods and is now showing signs of illness, such as vomiting, diarrhea, seizures, shaking, lethargy or weakness.

The Trend in Feeding Raw

Sushi may be chichi, but many people are also serving raw food to their pets because they believe it provides a healthier diet than commercial foods—not because they want to be culturally cool. This trend, however, is still highly controversial because of worries over salmonella, *E. coli* and other harmful bacteria, not to mention the fear associated with feeding bones.

Some nutritionists and veterinarians are strong advocates for a "bones and raw food" diet, also known as a "BARF" diet, saying that it is feeding your pets as nature intended. Always check with your veterinarian before making a change of this magnitude in diet.

Pros

- **Generally, cats fed raw foods are healthier.** Owners often report that their cats have shinier coats, healthier teeth, better breath, smaller stools and higher energy levels once switched to a raw diet. Raw diets also seem to benefit cats suffering from food allergies and digestive disorders.

- **Cats digest their meals better when fed raw food.** Chewing bones (which are pliable and less likely to splinter when raw) and eating raw animal parts requires more time, and this stimulates the release of more stomach acids, making the food easier to digest for the cat.

- **Chewing on bones and shredding raw meat cleans your cat's teeth.** Many raw advocates say there is no need for special cat toothbrushes and toothpaste if you feed raw. The shreds of meat act like dental floss and the bones provide a way to scrape off tartar and keep your cat's teeth clean.

- **Raw foods contain more water than commercially prepared foods.** A well-hydrated cat has fewer health problems. This fact especially applies to male cats, who often suffer from urinary tract blockages or infections.

- **By feeding raw, you know what your cat is eating.** There is real peace of mind knowing exactly what you are feeding your cat. No worries about poisonous additives or empty-calorie fillers in your cat's food if you are not relying on commercial foods to feed your cat.

Cons

- **Feeding raw may be time-consuming.** Preparing a raw-meal diet from scratch may take more time than dropping a few pieces of cat chow into a food bowl or opening a can of wet food.

- **You will have to add supplements to your cat's diet.** In order to ensure that your cat receives all the vitamins and minerals he requires, special supplements and vitamins must be included in his diet. Owners failing to put enough research into what supplements are needed to keep a cat strong and healthy has caused some cats to suffer serious nutritional deficiencies.

- **Some raw ingredients might be harmful.** There is evidence that raw fish for example will eventually produce a vitamin deficiency in your cat and can lead to seizure and even death.

- **Raw food must be refrigerated or frozen.** You will have to feed the raw food immediately or divide the meat into individual serving portions and freeze it in order to prevent spoiling.

- **Feeding raw can be costly.** Buying meats for your cat may cost you much more than you expected. Often raw enthusiasts purchase freezers to make feeding raw meat more economical because they buy meats on sale or in bulk.

- **Cleanup is messy.** Washing cutting boards and disposing of half-eaten raw animal parts is a turn-off for many would-be raw feeders.

Indoor Cat vs. Outdoor Cat

A debate rages as to whether a cat should be kept indoors for its entire life or allowed to go outside for periods of time.

There are pros and cons to both sides of the question, and potential pet owners should consider this issue before adopting a cat. For example, when you adopt an adult cat from a shelter or take in a feral cat off the streets, it may be difficult to keep him indoors if he is accustomed to coming and going freely.

If you adopt a cat that was formerly indoors-only and then thrust him into the outdoors, he can panic, run away or become sick. Young kittens are adaptable and will be happy as either indoor or outdoor cats.

Indoor Cats: The pros

- Keeping a cat indoors is safer than letting him roam. Outdoor cats can come in contact with fleas, ticks, worms and other parasites that pose health risks for your cat and your family.

- Indoor cats have a much lower risk of contracting a disease from neighborhood animals that are ill or ailing.

- The chances of an indoor cat being subjected to an attack from a stray dog, coyote or other predator are nil.

- Indoor cats avoid being poisoned by an angry neighbor or by someone leaving a toxic substance like antifreeze outside where a cat can get to it.

- Every year, cats are killed or maimed by seeking a warm spot in your car's engine area. An indoor cat is spared this fate.

- Male cats that are allowed to go outdoors are more likely to spray urine inside your home, because they come into contact with other male cats that spray and adopt the territorial habit.

- Thousands of cats are run over by cars every year, and your indoor cat is never going to be one of them.

The cons

- An indoor cat can feel caged, contained and trapped, which stresses his immune system and makes him susceptible to disease.

- Some pet owners feel that keeping a cat indoors borders on inhumane because it goes against the nature of this once-wild beast.

- Indoor cats can become depressed and lethargic.

- Indoor cats can become bored and overeat. Cat obesity is responsible for many diseases, including diabetes and heart disease.

- Indoor cats that started off as indoor-and-outdoor cats sometimes refuse to transition to only using a litter box. They will retain their waste until they get sick or urinate or defecate in inappropriate areas of the house.

- If using a cat for rodent control is important to you, it's good to note that indoors-only cats sometimes lose their ability to hunt, trap and kill their prey.

A compromise

Today many cat owners are allowing their cats out in protected areas such as screen houses, cat enclosures and closed patios, giving the cat a sense of the outdoors without exposing him to danger or disease. This way your cat gets to play on the grass, chase a butterfly and enjoy a sunbath while completely safe from danger.

Preventing Urinary Tract Infections in Male Cats

Urinary tract infections (UTIs) are one of the most common medical problems for male cats. Once your cat has had one UTI, he becomes prone to have another. As soon as you suspect that your cat has a symptom of a UTI, it is important to call your veterinarian immediately. Swift medical treatment can save him from severe pain or even death.

Symptoms

When a male cat develops too many crystals in the urethra, a dangerous obstruction may occur that prevents elimination of urine from the bladder. If this condition is not relieved within 48 hours, the male cat is at risk of dying from kidney failure because of the retention of toxins that were not removed by the kidneys. Contact your veterinarian if any of the following symptoms occur.

- Difficulty urinating
- Frequent and/or prolonged attempts to urinate
- Vocal sounds of distress like crying or howling
- Extreme licking of the genital area
- Uncharacteristically urinating in places other than the litter box
- Blood in the urine
- Less urine output in the litter box

Prevention

- **Make clean and fresh water available at all times.** With consistent water intake, your male cat's urinary tract is constantly being flushed. Urine is diluted, which helps to prevent blockage and infection.

- **Dump the dry food.** If you stop feeding dry food entirely, your cat will be forced to eat wet food, which introduces more water into your cat's system. In the wild, cats' prey is about 70 percent water; canned cat food has about the same percentage.

- **Feed a nutritious diet.** Your cat should be fed the best-quality food you can afford in order to ensure a healthy, nonstressed immune system. Do not have food available all day. Feed your cat smaller meals on a frequent basis to allow your cat's urinary tract to maintain a proper pH.

- **Resist "variety" feeding.** When you find products that your cat likes and that meet the "well-balanced" criteria, stick with them. Sometimes new cat food flavors actually stress your cat, weakening his ability to fight infection.

- **Offer extra-clean litter boxes.** Have one more litter box than you have cats. Keep a litter box on each floor of the house, including the cellar. By providing ample boxes, your cat never has to worry or retain urine because he's stuck somewhere without a box. Keep the boxes impeccably clean and in a safe, quiet area of the house. Studies have shown that there are more UTIs reported in cats that come from multiple-cat homes than single-cat homes. Territorial issues about litter boxes may keep the less-aggressive cat agitated and nervous about urinating, often making him retain his urine rather than face a fight at the box.

- **Minimize stress.** This is easier said than done. Our feline friends can't communicate what's bothering them, but often the same issues that cause humans stress equally affect the cat: moving to a new home; a loved one going away or dying; a new pet in the house; and family disharmony.

Preventing Constipation and Diarrhea

Constipation and diarrhea can negatively affect your cat's health. For example, over time, repeated cases of constipation may cause the colon to lose its muscular motility, causing a condition known as megacolon. Preventing these conditions is often easier than curing them. The following tips offer ways to avoid digestive problems in your cat.

Preventing constipation

- **Do** increase your cat's water intake. A lack of fluids will contribute to hardened stools, making them difficult to pass.

- **Don't** use any type of laxative unless prescribed by your veterinarian. Never use human enemas on cats.

- **Do** provide ample opportunities for your cat to exercise. Constipation is more common in cats that are inactive. Make an effort to play with your cat and encourage running, jumping and pouncing.

- **Don't** allow your cat's litter box to get dirty. Some cats are particularly picky when it comes to their box and refuse to go if it does not meet their standards, causing stool retention.

- **Do** feed the right foods. Wet food is more than two-thirds water and is best for your cat's digestive system. Seek the advice of your veterinarian about how to get added fiber into your cat's system.

Preventing diarrhea

- **Do** have your cat's stools checked by your veterinarian at least once a year. Cats and kittens are prone to a variety of intestinal parasites that are notorious for causing diarrhea. Often a deworming treatment will stop the diarrhea.

- **Don't** feed milk or most dairy products to your cat. Most cats are lactose intolerant and will develop runny stools and gas if given cow's milk to drink. Milk substitutes made specifically for cats are sold at grocery stores and pet shops, which give the milk-loving cat a delicious treat without starting intestinal upsets.

- **Do** change foods gradually. Feeding a new food too abruptly may cause an upset stomach. Start by mixing the new food with the old food, adding a bit more of the new food at each feeding until the transition is complete.

- **Don't** expose your cat to stress. Stress is a common cause of diarrhea in cats and therefore tends to appear when cats travel, move or are exposed to any uncomfortable event. Try your best to provide a peaceful and quiet setting for your feline friend to enjoy.

- **Do** ensure that your cat has access to clean, fresh water. Allowing your cat's water to sit for too long creates a breeding ground for bacteria. Bottled water or filtered water is usually a cat's first choice because of the lack of chemical smells. Put out ice cubes in your cat's dish when leaving kitty alone for a long period.

❖ **Do** change your cat's diet if diarrhea persists. Cats may benefit from special hypoallergenic cat foods if they are prone to food allergies, sensitivities or intolerances. Lowfat balanced diets, probiotics and small amounts of rice and plain yogurt all have helped firm cats' stools and calm intestinal irritations.

Common Plants That Are Poisonous to Cats

A surprising number of household and garden plants are poisonous to cats. Many cat owners don't even realize that the flowers they love or the bush they've just planted can contain dangerous toxins. Cats that ingest poisonous plants can suffer symptoms that include vomiting, diarrhea, seizures and even death.

More than 700 varieties of plants have been identified as containing toxins that are harmful to your cat. To protect your kitty, you should take an inventory of all plants, inside and out, and remove any that are potentially poisonous. Here's a quick guide to help you get started making your home cat-safe.

Indoor plants

They're meant to brighten up a home, but many common houseplants and flowers are deadly to your pet. Some of the most common culprits are amaryllis, aloe, elephant's ear, ferns, gardenia, hyacinth, iris, certain lilies and philodendron. Don't forget to take extra care at the holidays: Holly, mistletoe and poinsettia are all poisonous to cats.

Cut flowers may make beautiful accents for your house, but they can be toxic too. Bunches of carnations, daisies and tulips are as dangerous to cats as they are lovely to you.

Outdoor plants

If your cat goes outside at all, you need to take a hard look at your garden and backyard. Lurking in the bushes are some plants that can make your cat very sick. Some of the more common backyard plants you should be wary of include azalea, baby's breath, begonia, bird of paradise, branching ivy, buttercup, chrysanthemum, daffodil, foxglove, geranium, gladiolus, hibiscus, hydrangea, larkspur, morning glory, oleander, peony, primrose, sweet pea and wisteria.

Food plants

Some very common foods that you may have lying on the counter or even growing in your garden can also be dangerous for your cat. Some of the most common food plants you should be concerned about are apples, apricots, avocados, cherries, garlic, grapefruit, lemons, limes, onions, oranges, peaches and tomatoes. Keep your pet out of your garden and remember that even if the plants you have in your garden aren't toxic, the pesticides you use are.

What to do if your cat ingests a poisonous plant

If you think your cat has ingested a poisonous plant or any other toxic substance, you need to act quickly. Gather samples of the plant and call your vet or local emergency pet hospital. You can also call the ASPCA's Animal Poison Control Center. The telephone number is 888-426-4435. There is a $65 consultation fee for this service.

Sometimes the act of inducing vomiting can save your cat's life. You should have an emergency first-aid kit on hand that contains a bottle of 3 percent hydrogen peroxide and a large syringe to administer the peroxide to your cat orally. Some veterinarians recommend giving about 1 teaspoon of peroxide for each 10 pounds of your cat's body weight. Repeat the process no more than three times in intervals of 10 to 15 minutes.

Always check with your veterinarian before beginning this treatment or any other, since inducing vomiting is not recommended in some cases of poisoning.

For a complete list of plants that are poisonous to cats, visit the ASPCA's Animal Poison Control Center at *aspca.org*.

Cat Vomiting: A Sickness or a Lifestyle?

Cats have a reputation for frequent vomiting. Most of the time the cause of their digestive upsets is nothing to worry about. Their reputation for upchucking has become fodder for jokes about hair balls or vomiting cats. However, there's nothing funny about a cat vomiting if it's because he is suffering from an injury or a disease. Whether your cat is simply "tossing his cookies" or is truly ailing, you should be aware of the following information.

Prevention

Cats are very self-sufficient animals. After eating grass, acquiring hair balls or eating too much or too fast, cats use vomiting as a way to feel better immediately. It is common for some cats to vomit once or twice a week, especially if the cat has medium to long fur. Vomiting is the process that rids kitty of the furballs that he ingests from grooming.

- Brush your cat's coat every week to keep hair balls to a minimum.

- Because some cats are prone to motion sickness, they will vomit in the car or in the cat carrier. Try to position the cat carrier in a place where it won't wobble or slide during travel.

- When walking with the carrier, keep it from swinging up and down or side to side.

- Cats attempt to vomit up small objects they have swallowed, such as plastic toys, strings and rubber bands, to get rid of the obstruction. By cat-proofing your home, you can eliminate risking injury to your cat and reduce vomiting episodes by keeping small items away from kitty.

- Some cats eat their food too fast, then vomit it back up just as quickly. If you can't get kitty to slow down, feed him several smaller portions during the day.

- Sibling cats should be fed in separate areas to discourage gorging.

- Feed kitty on the floor so he doesn't have to jump down from a high area with a full stomach.

- Keep a list of what brand and flavor cat food you are feeding your cat. If vomiting occurs, mark down what food kitty consumed. You may find out, for example, that every time you feed your cat tuna, he has digestive problems. By keeping track of his foods, you can see if a specific cat food ingredient bothers his stomach, and you can eliminate it from his diet.

- Keep your cat safe indoors so he can't access dangerous poisons such as antifreeze, or eat foods such as chicken bones or rotting meats found in garbage cans.

- Many plants, both cultivated and wild, inside or outside your home, can be poisonous to cats. Check to see if any of your plants are poisonous to cats and get rid of the ones that are.

Intervention

A cat can become dehydrated quickly and an obstruction in his throat or bowels can cause serious illness. Seek medical help immediately if your cat exhibits any of these symptoms.

- He has thrown up numerous times over an eight-hour period.

- There is blood in his vomit.

- He is vomiting and not drinking water.

- He tries to eat but regurgitates his food instantly.

- He has signs of sickness, such as weakness, coughing, shaking or diarrhea.

Pet Care 101: Grooming Your Cat from Nails to Tail

Every time you look, your cat is grooming himself—actually, almost three hours a day, when he's not napping, that is. So, do you really have to groom him, too?

There are a few basic grooming techniques that should be made part of every cat owner's weekly routine. Here's a quick guide to help you start grooming like a pro.

Make grooming enjoyable

The best time to begin grooming your cat is when she's a kitten. But if she's an adult, you can still make grooming a good experience for both of you. Start by approaching your cat when she's in a restive mood. Begin by petting her, stroking her and speaking to her softly. You may even want to groom her with your fingers first so she gets used to the feeling. Then bring out the comb and gently pull it through her fur. If you already know she has tangles or matted fur, only comb her where you can glide the comb easily; you don't want her first real grooming experience to be painful. You may even comb her for only one or two minutes the first time; that's OK. When you're finished with the session, praise your feline and even give her a treat.

Bathing

If your cat has gotten something on her coat or it is excessively oily, it's time for a bath. Place a rubber mat in the sink or tub to give her a nonslip surface to stand on. Fill the sink or tub with a few inches of lukewarm water. Let your cat hook her paws over the edge of the sink or tub so she doesn't feel trapped. Use a medicated baby shampoo or special animal shampoo and massage the shampoo gently into her coat. Rinse kitty by using a cup rather than a spray hose. Avoid getting water and soap in her eyes, ears and nose. Dry her with a soft towel until she is damp-dry.

Brushing

Grooming isn't only for the sake of appearance. It prevents matting (even in short-haired cats) that can irritate your cat's skin, leading to itching and infection. Brushing and combing your cat also gives you the chance to check for ticks and fleas. In fact, grooming is a form of preventive pet care. While grooming, you may discover a lump in your cat's skin that requires immediate veterinary diagnosis. Finally, if you and your cat get to the point where grooming isn't troublesome, it gives you some time to spend together.

Brushing your cat doesn't just give her a healthy, silky coat; it helps keep her from getting hair balls and it's a great way for the two of you to bond. Start when your cat is relaxed or sleepy. Use a hard metal comb and work out any mats. Use the comb's wide teeth for long-haired cats such as Persians. A slicker brush with a slanted handle allows you to get down to the bottom layers of the fur to comb out loose or dead hair and to remove debris and dander. For a comparison of the many different types of grooming tools available for cat pet care, see the "Articles" section at *drsfostersmith.com*. Be gentle, especially in sensitive areas like the belly. Switch to a bristle brush to help remove the loose hair. At first, keep your brushing sessions short and stop when your kitty lets you know she's had enough. Long-haired cats require more brushing than their short-haired cousins. Longhairs require daily brushing, while once a week is enough for most other breeds.

Eyes and ears

If your cat's eyes have a crusty buildup, it can clog her tear ducts, which can lead to conjunctivitis or pinkeye. Simply use a moist cotton ball and gently wipe outward from the inner corner of the eye. If the eyes are persistently weepy, see a veterinarian.

To clean your cat's ears, gently fold them back and swab with a cotton ball moistened with ear-cleaning solution. Do not scrub the ears or put any objects or liquid into the ear canal itself.

Teeth

Pets need dental care just as humans do. Help your pet get used to the idea of having her teeth brushed by first massaging her gums. Add a bit of specially formulated pet toothpaste to your finger so she can grow accustomed to the taste. Then apply some of the paste to a pet toothbrush to let her get used to the texture of the brush. When she's ready, lift her lip and gently begin to brush. Be sure to praise and soothe her. Start with just a few teeth, then add in more teeth as her tolerance to the brushing increases.

For more information about cat dental health, see page 90.

Nails

If left unattended, your cat's nails can become ingrown and infected. Get your cat used to having her paws held and her claws manually extended by starting the process when she is a kitten. Gently hold her paw and press the bottom of the pad to extend the claw. Use special cat nail trimmers and be sure not to cut off too much nail or you might nick the quick or vein. Never cut into the red part of the nail. If you're unsure or your cat is too unwilling, take her to a grooming service.

Seek help from a pro

Sometimes your cat's matted fur may be so thick that you can't safely remove it yourself. In that case, bring your cat to a professional groomer, who will shave the problem areas. Whatever you do, don't cut your cat's knots out with scissors or other sharp instruments. Cats' skin is very thin and you can easily cut your cat, leaving a gaping hole that requires stitches.

Ridding Kitty of Fleas and Parasites

Getting rid of kitty's fleas and other parasites can be a lot like getting rid of annoying relatives: They keep coming back. But unlike relatives, fleas can be banished for good.

If you are patient, you can rid your cat and your home of unwanted parasites with a little work and a few antiparasite products.

Bathe your cat whether she likes it or not

Use a good-quality but gentle flea shampoo. Place your cat in a prefilled tub containing two or three inches of warm water. Use a small amount of shampoo on the cat's body, but be very careful once you reach her head. Do not use flea shampoo on her face. Rinse her with cupfuls of clean, warm water. If your cat is not used to bathing or is very skittish, you can take her to a professional groomer to be flea dipped.

Groom kitty after bathing

Once your cat is fully dry, you can begin combing her with a fine-toothed metal flea comb to snag any stragglers. Gently run the comb through her fur and look for signs of fleas or ticks. If you come across ticks, remove them by tweezing the tick until it lets go. Drown it in a bowl of water or flush it down the toilet.

If you find fleas on your cat, masking tape is a good way to collect the ones that stick to the comb. Simply put the tape on the comb and the fleas will transfer to the tape. Or fill a bowl with hot, soapy water and dunk the comb into the water to drown the trapped fleas. Sometimes bathing and combing won't contain or kill all the fleas, and you will need to apply a flea-and-tick-control product between kitty's shoulders to kill off future generations. Products such as Frontline are very effective in eliminating both fleas and ticks, while natural products usually take more applications to work.

Launder the cat's bedding

After kitty is clean, you must launder all cat bedding in hot water in your washing machine. Include your bedding if she snuggles in your bed at night.

The flea's outer shell is coated in a waxy substance that keeps the fleas healthy and hydrated. Fifteen minutes in hot, soapy water and the shell will soften, eventually killing the flea.

Vacuum the carpets

This should be done weekly or more often if you can. Once you've vacuumed the rugs, around bedding, cushions and anywhere else the cat hangs out, take the vacuum bag to the outside trash can and throw it away. Treat your cat's favorite spots and around the litter box with a flea spray, but check with your veterinarian to find out her recommendations as to what product to use.

TIP: Try the flea spray in an unobtrusive spot first in case it affects the color of the carpet. Don't forget to clean the less-obvious areas, such as behind doors, in closets, in corners of rooms and any warm, moist spot.

With vigilance and a step-by-step approach, you can rid your home and your cat of most parasite infestations. If you find that the fleas have won the battle and have taken over your home, it's time to call in a licensed exterminator to help you reclaim your pest-free home once again.

The Best Way to Pill a Cat

As many cat owners know, giving a cat a pill can be a difficult task. However, common feline health conditions often require oral medications for treatment, making pill-giving a necessary evil. Read on for the best way to pill a cat, and tips to make the process less painful for everyone involved.

- **Hide the pill in food.** Cats are intelligent creatures and finicky eaters. Most cats simply eat around a pill hidden in their food. However, this method is worth a try because it involves the least amount of stress for both cat and owner. Try hiding the pill in a soft cat treat or a small piece of cheese or hot dog. Watch to make sure that your cat actually consumes the pill and doesn't spit it back up.

- **Ask for help.** With practice (and a relatively docile cat), giving a cat a pill can be a one-man job; however, in most cases, having an extra set of hands helps immensely. Recruit a friend or family member to help hold your cat still while you administer the pill.

- **Insert the pill.** The moment of truth: Balance the pill on your forefinger and slide it into the cat's mouth by forcing her jaw open gently at the side of her mouth. Quickly deposit the pill as far back as you can, then gently hold her mouth closed and keep her head up for a few moments for her to swallow. If it's your first time, make sure the vet gives you a good demonstration before you leave the doctor's office. There is excellent video help available from the Cornell University College of Veterinary Medicine (*partnersah.vet.cornell.edu/pet/fhc/pill_or_capsule*).

- **Restrain your cat properly.** The best way to restrain a cat while giving her a pill is from behind, wrapping a towel around her for your own protection. Position your cat on a slippery, elevated surface, which makes your job easier and limits her chance of escape. Use your upper body to cradle the cat from behind and firmly hold her forelegs near the elbow joint (this is where the extra hands are helpful).

- **Invest in a pill gun.** Whether you have help or you're working solo, giving a cat a pill is much easier when you use a pill gun. This device, which works like a syringe, is a good alternative for cats that bite or are resistant to swallowing a pill. Using a pill gun means you can get the pill positioned closer to the back of your cat's throat without having to risk being bitten. Even gentle cats are more likely to bite when they're being held down and something is being jammed down their throats. Pill guns can be purchased at your local pet supply store or from your veterinarian.

- **Make sure the cat swallows.** Getting the pill in your cat's mouth is just half the battle. Making sure she swallows the pill is another story altogether. Start by lubricating the pill with some butter, which makes it slide more easily down her throat (and also makes her more willing to swallow it). Regardless of whether you use a pill gun or your fingers to administer the pill, you can use the following technique to help ensure that she swallows it. Once the pill is on her tongue, gently hold your cat's mouth closed and stimulate swallowing by massaging her nose or the front of her neck with your fingers.

One Fat Cat: Cat Obesity

As people grow fatter and fatter, so do our pets. Almost half the cats in America are considered obese or weigh 20 percent more than their ideal weight. Cats are becoming overweight for the same reasons people are: too much to eat, too much junk food, too little exercise and eating out of boredom.

Diseases that are prevalent in overweight people are now becoming common in cats. An overweight cat is at risk for diabetes, arthritis and hepatic lipidosis (fatty liver disease) in the future. Heart disease, skin problems, breathing problems and depression can occur when a cat is obese and inactive.

It is up to the cat's owner to start kitty on a diet before his obesity causes a serious health issue. Begin by making a veterinary appointment to ensure that your cat has no other health issues and is ready for a restricted-calorie diet. Your veterinarian can help you put kitty on a program of gradual weight loss and increased activity to bring your cat back into a normal weight range.

How cats are fed It is unnatural for cats in the wild to have food available at all times. Kitty won't get slim if you have a self-feeder or leave food out all day. A good first step in battling cat obesity is to change this particular feeding habit. Instead, feed two to four small meals per day of a high-quality cat food and don't attempt to starve your cat for instant results, since this could cause serious health problems.

Carnivores eating carbohydrates Cats have very little amylase (carbohydrate-digesting enzyme) in their pancreas and none in their saliva. Cats were not meant to eat carbohydrates, yet many cat foods contain carbohydrates that act as fillers. Dry cat food often has sugar and flour added to keep the chow from breaking apart. The added carbohydrates fatten your cat without adding nutrition. Start your cat's diet by changing to a high-protein, high-moisture, low-carb cat food to boost nutrition while eliminating useless bulk.

Feeding kitty treats Most cat treats have even higher carbohydrate levels than commercial cat food. Your cat does not need the empty calories found in manufactured treats. Stick to a planned feeding schedule, and if you must give treats to your cat, prepare cooked fish or chicken cut into tiny bites. Offer these protein-rich tidbits rather than others that are high in sugars and carbs.

Exercise makes a difference A fat cat loses the desire to run, jump and play, yet exercise is important in helping him lose his extra weight. Start by purchasing or making toys that you can throw, drag or roll. Engage your cat in play by talking to him and showing excitement about the toys. Schedule five to 10 minutes of playtime three or four times a day.

Are you thinking about expanding your cat family? If so, a new playmate is a wonderful way to get your obese cat running, jumping, pouncing and exercising.

How to Exercise Your Cat

There's more than one way to help your kitty work out. First, motivate her to exercise by putting her food and water dishes farther away from the litter box. Second, train your cat to eat out of a Play-N-Treat ball (available at pet stores), which has a hole that spills out kibble when it's batted. Then hide it behind doors, under the bed and so on, so she has to hunt for it. Finally, tempt your cat with Ping-Pong balls, paper bags, a mouse toy or a feather at the end of a stick.

Cat Dental Health

The Cheshire cat is known for his large grin, but if he had swollen gums, loose teeth and cavities, he wouldn't be smiling at all. The health and care of a cat's teeth can be easily overlooked, but dental care is as important to a cat's well-being as it is for humans.

The American Veterinary Dental Society reports that approximately 70 percent of cats exhibit signs of dental disease by the time they are 3 years old. Dental disease can be painful for your cat, and it can progress quickly into more serious physical ailments, such as problems with kidneys, heart and lungs.

Periodontal disease

If you haven't been diligent in caring for your cat's teeth, you may notice she has bad breath. This may be a symptom of a gum infection called periodontal disease. If you notice her having trouble eating dried cat food, her teeth might be loosening to the point that her chewing is impaired. If your cat has other symptoms of gum disease such as bleeding or swollen gums, see your veterinarian before she needs numerous expensive extractions or becomes physically ill from the disease.

Dental procedures

At your cat's annual veterinary checkup, the veterinarian should also examine your cat's teeth. Your vet should check for oral development, plaque buildup, anomalies, swelling and tumors. If the removal of tartar or a more thorough exam is necessary, your cat might need a mild anesthetic.

With your cat under anesthesia, your vet can:

- Perform a more thorough examination of her teeth, throat, tongue and gums.

- Take oral X-rays as necessary. This will help the veterinarian locate any anomalies that are not visible to the eye.

- Scale your cat's teeth. The veterinarian will remove the buildup of tartar and polish the teeth, which smooths any rough surfaces.

- Use an antiplaque substance, such as fluoride, or apply a sealant to help prevent decay.

Dental care

As the cat owner, you have a responsibility to care for your cat's teeth on a daily basis. Start a brushing schedule when your cat is a kitten, so she is used to the procedure throughout her life.

Make your kitten or cat comfortable with brushing by dipping your finger into the liquid found in canned tuna. Rub around her gums, mainly at the area where the teeth and gums meet. Next move on to using a piece of gauze over your finger and rub the area with a circular motion. Finally, use a brush designed for cats and a no-rinse pet toothpaste.

Instead of soft treats, give your cat hard treats, which will remove plaque from her teeth. A quality high-protein dry food, given in small amounts, will also help remove plaque. Most important, make a check of your cat's teeth periodically to look for brown teeth, gum swelling, pus, loose teeth or redness. If any of these signs are visible, make an appointment for a veterinary exam so your vet can assess her dental health and help fix any tooth or gum problems she has before the problem gets worse.

Summer Pet Safety Guide

Sunburn

Like their owners, animals can also get a sunburn. And light-colored cats are at special risk for skin cancer. To keep burns at bay, apply pet-safe SPF 15 or 40 sunscreen (found at pet stores) to the tips of both cats' ears. Cats often try to wipe off creams, so you may need to use a flea spray with SPF instead.

Heatstroke

Cats don't perspire; they release body heat by panting and through the pads of their feet, so they have a harder time cooling down. Limit their time outdoors and never leave them in a car—it can reach over 100 degrees in minutes. Make sure they have plenty of shade and fresh water, and keep your home cool. Be careful with snub-nosed breeds (Persian and Himalayan cats); they have a harder time panting. If your pet displays signs of heat stress—heavy panting, rapid pulse, vomiting, lethargy—lower his body temperature immediately by applying cool, wet towels, and call your vet.

Outdoor Poisons

Fertilizers, herbicides and insecticides All three of these common garden products can spell disaster if pets chew into the packages. Insecticides in particular can be fatal if eaten. Store concentrated products somewhere inaccessible to pets. If your pet ingests toxins, call Animal Poison Control immediately (888-426-4435). If he's gasping or seizing, rush him to the vet.

Antifreeze This substance tastes sweet to animals, yet it's anything but. "Kidney failure can develop within hours of ingesting it," says Steven Hansen, DVM, director of the ASPCA Animal Poison Control Center. Use less-toxic antifreeze made with propylene glycol, not ethylene glycol (the label says "dog-safe"); store it securely and watch for car leaks. If your pet ingests antifreeze, take him to the vet right away.

Fleas and ticks A multitude of flea bites can potentially lead to serious allergic reactions and skin problems. Ticks can transmit Lyme disease and Rocky Mountain spotted fever, both of which cause joint pain and neurological problems in animals.

To fend off fleas, forget flea collars: They're not that effective. Your best bet? Do a daily comb-through and get rid of any you find with flea-and-tick shampoo (get your vet's OK on the brand first). If you see a tick, don't shampoo; instead, gently pull it straight out with tweezers. As a preventive measure, apply a flea-and-tick solution each month. But never use one made for dogs on cats: They contain permethrin, which can be fatal to felines.

Traveling with Your Pet:
5 Essential Pet Health Records

Before you pack up your car for a summer road trip with your cat, be sure your pet's health records are handy. These important documents will prove vital in the face of travel-related health hazards, such as increased exposure to fleas and ticks, a medical emergency or an unexplained illness. Here is what your pet health records should include:

❖ **Provide your veterinarian's contact information.** The first page of your pet health records should include your veterinarian's name, address, phone number, e-mail address and fax number. This information will come in handy if you need immediate advice on how to treat an injury or if another vet needs more information on your pet's medical history.

❖ **Keep records on shots and vaccinations.** If your pet is exposed to a dangerous disease during travel, you must inform the treating vet of his shots and vaccinations, difficult details to remember, especially in an emergency. Have a copy of your pet's documented shot and vaccination history. This information will give the vet an overview of your pet's health and help him decide how to treat him.

❖ **Supply written prescriptions for medications.** Does your pet take a daily medication to control an illness or condition? If so, don't forget to bring a copy of the written prescription with you on the road. Medication can get lost in the shuffle during travel, so you may need to get a refill. Imagine how your trip will be ruined if this isn't possible.

❖ **Document past medical emergencies.** On a piece of paper, list your pet's past medical emergencies, and keep them with his health records. Provide the attending vet information on visits that required surgery, long-term medication or a change in lifestyle so he can make appropriate care decisions based on your pet's medical history.

❖ **Discuss current diet and allergic reactions.** Does your pet suffer from allergies or swell up if he eats chicken? You need to list this type of information in your pet's health records. It is particularly helpful when you stay in a five-star hotel, and hotel staff is planning the menu for your pet. Those people need to know what not to feed your pet so he stays healthy.

Keeping these five vital pet health documents handy is essential for the well-being of your cat or dog as he travels. Be sure to keep all original documents at home and put a copy in a packet or folder that you can take with you. With hope, you will never need to use any of this information, but if you do, you'll be adequately prepared.

While-You're-Away Pet Care Checklist

When you must leave home without them, you need to provide more pet care for your cat or dog than someone merely coming to fill the food bowl or scoop the litter. Here is a checklist for what needs doing before you leave.

94

- **Have your pet meet her sitter in advance.** Always arrange a meeting with the pet sitter and your animals before you leave town, especially when it's the first time you're using this person. This increases the chances that your cat will be comfortable with the pet sitter, and that you can fully discuss your pet's needs.

- **Nail down your departure and arrival times with the sitter.** If there are any changes while you're away, call your sitter right away, especially if your trip takes longer than expected or you miss a flight home.

- **Leave all your phone numbers with the sitter.** Do this even if you think your sitter already has them. Leave your home and cell phone numbers, your veterinarian's number, and the phone number of someone who can fill in for you in case you can't be reached.

- **Decide beforehand how often you want to speak with the sitter.** If you want to call for a moment every day, let the sitter know. Don't be embarrassed to express your need for reassurance that your dog or cat is doing OK in your absence.

- **Don't forget to give your sitter your house key.** Also, tell your sitter who else has your key—and their phone numbers—in case the sitter loses the key.

- **Make sure your pet's ID information is up-to-date.** This includes tags, microchips, etc. This will be infinitely helpful in case your pet gets lost. This is one of the most important aspects of pet care.

- **Tell your sitter if your pet is not allowed somewhere in the house.** This can mean everything from the basement or the kitchen to whether your pet is allowed on the bed or sofa. Cats can be quite clever at trying to get away with something when their owner isn't there to scold them.

- **Write a complete list of everything your pet needs.** This includes: >> Food and where it's kept, and how much and how often your pet is fed. >> Medicines, including where they're kept and written instructions on how they're administered. >> Your pet's favorite toys. >> Whether any other animals are allowed near your pet.

- **Leave a copy of your pet's medical records with your sitter.** Also leave the name and number of an emergency veterinary hospital you would use if your regular veterinarian is not available.

- **Leave a signed and dated note giving permission for your sitter to take your pet to your veterinarian in an emergency.** Discuss beforehand under what circumstances you want to be notified first and what constitutes a real emergency in which your sitter should take your pet off to the vet immediately. And—this is very important— discuss how you will pay for emergency veterinary care. Some animal hospitals won't treat an animal unless they have a credit card number to bill or receive advance payment.

- **Tell your sitter your pet's favorite hiding places in the house.** Cats will often hide when an unknown sitter comes for the first few times. Also, cats tend to hide when they're sick, so it's important that your sitter be able to find your cat.

Notes